The CFO's Operating Manual

Cash, Capital, and Strategic Finance

Joseph Trek

Published by NobleTrex Press

For permissions and other inquiries, write to:

P.O. Box 3132, Framingham, MA 01701, USA

Contents

Introduction

The room is always quietest when the numbers stop adding up.

You know the feeling. It is Tuesday morning, the third week of the quarter. The CEO is pacing in front of a whiteboard covered in exponential curves that represent "hope." The Head of Sales is explaining why a pipeline of "verbal commitments" is as good as signed contracts. The Head of Product is asking for three more engineers to build a feature that doesn't have a revenue model yet.

And then, every head turns to you.

They aren't looking for enthusiasm. They aren't looking for a vision. They are looking for the ground beneath their feet. They want to know if the runway is long enough for the plane to take off, or if they are all just speeding toward a cliff in a very expensive car.

This is the paradox of the modern Chief Financial Officer. You are expected to be the catalyst for growth and the guardian of the brakes. You must be the strategic visionary who sees five years into the future, yet also the tactical operator who knows exactly why the bank reconciliation didn't clear yesterday. You are the only person in the room who has to hold two opposing thoughts in your head at the same time: the limitless potential of the business and the brutal, finite reality of the bank account.

Welcome to the loneliest, most exhilarating seat at the table.

If you picked up this book, it's likely because you've realized that the standard finance curriculum didn't tell you the whole story. Business schools are excellent at teaching you how to calculate the Weighted Average Cost of Capital (WACC) in a vacuum. They are terrible at teaching you what to do when your largest customer stretches their payment terms to 90 days right before your debt covenants are tested.

Textbooks teach you GAAP accounting; they rarely teach you how to tell a founder that their "investment mode" is actually just a cash bonfire.

This manual is written for the gap between the theory of finance and the messy, chaotic reality of operating a business. It is about the machinery of money-how to build it, how to fuel it, how to steer it, and how to keep it from blowing up.

The Physics of Business

There is a misconception that finance is about math. It isn't. If finance were just about math, we would have replaced CFOs with spreadsheets in 1995.

Finance is the physics of business. It is the study of forces. We look at the energy intended for motion (Capital), the friction that slows it down (Costs), and the distance it can travel (Runway). Just as an engineer cannot cheat gravity, a business cannot cheat the economic reality of its unit economics forever. You can defy gravity for a moment with enough thrust-venture capital or cheap debt-but eventually, the engine has to sustain the flight.

Your job is to understand these forces better than anyone else. While the rest of the C-suite speaks in the language of *narrative*-stories about brand, culture, and disruption-you speak the language of *truth*.

That truth begins with the mandate. Before we touch a single ledger entry, we have to understand what we are actually doing here. We are not just scorekeepers. The "Scorekeeper CFO" is a relic of the 20th century, a person who sat in the back office and told you what happened three weeks ago. Today, you are an architect of value. We start our discussion by tearing down the old definition of the role and rebuilding it around stewardship and fiduciary duty. We have to look at the business model itself-the raw mechanics of how the company makes a dollar-and the mathematical backbone of finance: the time value of money. This isn't abstract theory; it is the realization that a dollar today is a tool, while a dollar next year is merely a promise.

The Map and the Territory

Once we understand the physics, we need a dashboard. This is where accounting comes in, but not the way you learned it in Intro to Financial Accounting.

Accounting is often dismissed as the boring part of the job, the realm of compliance and dusty binders. This is a fatal error. Accounting is the operating system of the enterprise. If the data going in is garbage, the strategic decisions coming out are hallucinations.

We are going to demystify the "dark arts" of the three key statements. The Income Statement is a story of effort and reward. The Balance Sheet is a snapshot of health and history. But the Cash Flow statement? That is the lie detector. You can massage net income with clever accruals and capitalization policies. You can make EBITDA look like whatever you want it to look like. But you cannot fake the cash balance. It is binary. You either have the money to make payroll, or you don't.

We will walk through how to build a reporting engine that generates trust. Speed matters, but accuracy matters more. A CFO who is wrong twice loses the room forever. We will discuss the "Close Process"-not as a monthly chore, but as a high-performance ritual that turns chaos into clarity.

Oxygen and Survival

Whatever your strategy is, it dies without liquidity.

Cash is the oxygen of the business. You don't wake up thinking about oxygen, but if it disappears, it is suddenly the *only* thing you think about. We are going to treat cash management not as a back-office treasury function, but as a survival skill.

I have seen profitable companies go bankrupt. It happens more often than you think. They grew too fast, bought too much inventory, or signed a lease they couldn't sustain during a downturn. They looked great on the P&L, but they suffocated because their cash conversion cycle was broken. We are going to open the hood on Working Capital-the gears of receivables, payables, and inventory that determine whether your growth consumes cash or generates it.

This leads us to the most critical tool in your arsenal: the 13-Week Cash Forecast. If you take only one thing from this manual, let it be the obsession with short-term liquidity visibility. Knowing where your cash will be in 90 days allows you to sleep at night. Not knowing is a dereliction of duty.

Predicting the Unpredictable

Once we have secured survival, we earn the right to look forward. This is where the CFO transitions from mechanic to navigator.

Planning and performance management-often called FP&A-is usually done wrong. Most companies treat budgeting as a political negotiation where department heads inflate their requests and finance cuts them by 10%. The result is a budget that everyone agrees to but nobody believes in.

We are going to throw out the static annual budget and replace it with dynamic planning. The world moves too fast for a plan that was locked in last November to be relevant next June. We will explore rolling forecasts and driver-based modeling. We stop asking "How much did we spend?" and start asking "What are the unit economics of our growth?"

This section is about driving accountability. It is about the uncomfortable conversations where you have to point out that the marketing spend isn't yielding the promised ROI, or that the operational efficiency gains were a mirage. You are the one who has to hold the mirror up to the business, even when the reflection is ugly.

ngine

ur ambitions will exceed your resources. You will need al.

This is where the game changes. Stepping into the capital markets-whether dealing with a commercial banker for a line of credit or pitching a venture capitalist for Series B equity-requires a different demeanor. You are selling a piece of the future.

We will break down the cost of capital, not just as a formula, but as a concept of "hurdle rates." Every dollar you take from an investor comes with strings attached-expectations of return, loss of control, and covenants. We will discuss how to design a capital structure that fuels growth without handing over the keys to the castle. Debt is a power tool; used correctly, it amplifies your returns. Used carelessly, it will dismember you.

We will also look at the human side of fundraising. Investors and lenders are not algorithms. They are people assessing risk. Your ability to communicate a coherent, credible equity story is just as important as your spreadsheet.

Allocating the Chips

With capital in hand, the question becomes: Where do we put it?

Capital allocation is the CEO's most important job, but the CFO provides the map. Do we invest in R&D? Do we hire more sales reps? Do we buy a competitor? Or do we simply return the money to shareholders?

We will explore valuation and M&A. Mergers are exciting. They are headline news. They are also statistically likely to destroy value. The "synergies" that look so beautiful in a slide deck have a nasty habit of evaporating when human beings get involved. We will look at how to strip the emotion out of deal-making and focus on rigorous diligence. We will learn how to value a business based on its intrinsic cash-generating ability, not just what the market hype says it's worth.

The Shield and The Architect

Finally, we must address the things that can kill you from the blind side.

You are the Chief Risk Officer. You are the person who asks, "What if the currency collapses?" "What if our data center floods?" "What if our biggest supplier goes under?" This is the "Shield" function of the CFO. It includes the unsexy but vital worlds of insurance, tax compliance, and internal controls. It's about building a fortress that can withstand shocks.

But you cannot do this alone. The days of the CFO as a lone wolf with a calculator are over. You need to build a machine. We will close by looking at the "How"-the systems, the data architecture, and the team. The modern finance function is a tech stack. If you are still manually copy-pasting data from one CSV file to another, you are failing. We need to automate the rote work so your team can focus on the strategic work.

A Manual, Not a Bible

Finance is not a religion. There are very few absolute truths. There are principles, there are frameworks, and then there is judgment.

This book is called an *Operating Manual* for a reason. It is not meant to sit on a shelf and look impressive. It is meant to be used. It is designed to be pulled out when you are designing a board deck, when you are arguing with a banker about covenants, or when you are trying to figure out why your gross margin just dipped three points.

As you read, remember that the goal of the CFO is not to be the most popular person in the company. It is to be the most trusted. You are the steward of the scarcity that gives value to everything else. In a world of infinite ideas, you are the master of finite resources.

So, pour another cup of coffee. Open the spreadsheet. Let's get to work.

Chapter 1

The CFO Mandate: Value, Strategy, and Foundations

If a company were a ship, the CEO would be the captain pointing toward the horizon, and the CFO would be the chief engineer and navigator combined. It is not enough to simply count the cargo; you must ensure the engines are efficient, the fuel will last, and the route is economically viable. In this chapter, we strip away the jargon to reveal the three pillars of financial leadership: the mandate to protect and grow value, the mechanics of your business model, and the undeniable force of financial gravity known as the Time Value of Money.

1.1 Defining the CFO Role: From Bean Counter to Co-Pilot

Picture the moment that makes a CFO famous inside a company—not on CNBC, not in the annual report, but in the hallway right after the meeting. The CEO has just pitched a bold move: acquire a competitor, double down on a new product line, lock in a long-term supply agreement, open three countries at once. People are energized. Then someone turns to you with that look: *Are we doing this or not?* In the old caricature, this is where finance kills the vibe with a spreadsheet and a lecture about budget. In the modern version, this is where you earn your seat as co-pilot: you keep the plane out of the mountain *and* you help pick the altitude.

The CFO job is not "make the numbers look good." It's "make the numbers true, make the business solvent, and make the future fundable." You're the executive who has to live in two time zones at once. One foot is planted in obligations: payroll clears on time, taxes are filed correctly, covenants are respected, controls prevent fraud, the financial statements don't accidentally mislead anyone. The other foot is planted in possibility: which bets to place, which to avoid, how to sequence them, and how to pay for them without mortgaging the company's resilience.

That tension—guardian and growth partner—isn't a branding exercise. It's physics. A company can be strategically brilliant and still die from cash starvation, a bad capital structure, or a credibility gap with lenders and investors. A company can also be impeccably controlled and still lose the market because it refused to invest until certainty arrived (which is another way of saying it never invested). The CFO sits at the intersection of those failure modes. Your value is not that you can say "no." It's that you can say "yes, if," and make the "if" concrete enough that adults can make a decision.

The "bean counter" label persists because it has a grain of truth: the CFO is one of the few roles where being boring is sometimes the point. Nothing is more underrated than a company that pays people correctly, closes the books without drama, and produces numbers that leadership trusts even when they're ugly. Trust is a compounding asset. Once people believe you won't sugarcoat, they listen when you advocate for risk. That's the paradox: the CFO earns permission to be bold by being relentlessly honest about what's real.

The role evolved because the business environment evolved. When capital was cheaper, growth was more forgiving, and cycles felt slower, finance could play referee and historian. But modern companies run faster, with more external scrutiny and more internal complexity: subscriptions, usage-based pricing, multi-entity tax structures, global supply chains, security threats, and stakeholders who want clarity on everything from margins to ethics. In that world, the CFO becomes the executive who translates strategy into commitments: commitments to spend, commitments to hire, commitments to customers, commitments to lenders, commitments to regulators. Translation is not clerical work. It is leadership under constraints.

A modern CFO is measured less by what they *know* and more by what they can *align**. You align incentives (so teams don't "win" by losing money), align timing (so cash arrives before bills), align the story (so the board and investors understand why your plan is rational), and align the operating system (so decisions can be made with the same numbers, not a dozen competing truths). In other words, the CFO becomes the company's chief integrator: the person who makes the business legible enough to steer.

The CFO role also changed because the CEO role changed. CEOs are expected to be storytellers, culture-setters, and strategic allocators. They're pulled into product, talent, partnerships, brand, geopolitics, and crisis response. They need a counterpart who can turn ambition into a plan that survives contact with payroll dates and debt maturities. The CFO becomes the CEO's most pragmatic partner: not the one who dampens ambition, but the one who turns it into a sequence of fundable steps.

This is where the co-pilot metaphor earns its keep. A pilot needs instruments because humans are bad at sensing speed, altitude, and drift in clouds. Companies have their own clouds: a quarter with great bookings and terrible cash, a growth spurt that hides churn, a hiring wave that buries productivity, a product launch that looks successful while it quietly compresses gross margin. The CFO is the instrument panel. Not the person who flies the plane alone—the CEO and leadership team still make the calls—but the person who makes the real dynamics visible early enough to correct.

There's an uncomfortable truth embedded in that: many organizations *don't want* visibility. They want comfort. They want forecasts that don't embarrass anyone, KPIs that always go up, and plans that feel inevitable. The CFO is often the first executive asked to participate in a polite fiction: "Can we present this as a one-time issue? Can we exclude that cost? Can we call this investment 'non-recurring' again?" A co-pilot can't do that. You can't manage altitude by editing the altimeter. Your credibility depends on treating reality as a partner, not an opponent.

The fiduciary side of the job gets summarized as "controls" and "compliance," which makes it sound like paperwork. It's actually a set of promises. You promise employees they'll be paid. You promise customers you'll deliver what you sell. You promise suppliers you're good for the check. You promise lenders you'll follow the rules attached to their money. You promise shareholders and regulators that you'll report performance without deception. These are moral promises, but they're also economic ones: once you break them, the cost of capital rises, talent leaves, and your strategic options narrow.

Controls exist for the same reason guardrails exist on mountain roads: not because you plan to crash, but because you recognize that humans and systems are imperfect. The CFO builds those guardrails through segregation of duties, approval workflows, reconciliations, and audit trails—yes, the boring stuff. But the strategic payoff is enormous. A company with strong controls can move faster because it doesn't spend its life cleaning up preventable messes. Speed is not the opposite of control; speed is what control buys you.

Risk management is often framed as fear. In practice, it's about *survivability*. The CFO's most strategic act might be ensuring the company can live through a bad year without begging for mercy. Liquidity is not a trophy; it's a shield. When you have a cash buffer, you can invest in a downturn, keep talent, and negotiate from strength. When you don't, you accept terms you hate, cut muscle, and sell the future to pay for the past. A CFO who treats liquidity as "idle cash" misses the point. Liquidity is optionality.

Now for the part that surprises first-time CFOs: being a guardian doesn't mean being conservative. It means being *explicit*. Guardianship is demanding that risks are named, priced, and monitored. It's saying, "We can take this risk, but we will size it, stage it, and attach triggers." The company that blows up is rarely the one that took risks. It's the one that took risks unconsciously, without realizing they were compounding.

Consider a simple example. Sales wants to offer customers net-90 terms to win deals. That's not just a "collections" issue. It's a financing decision. You're lending your customer money for three months. If your gross margin is thin, or your vendor terms are net-30, you've just designed a cash trap. The CFO's job is not to scold sales for being aggressive; it's to quantify the cash cost of that aggression, decide whether it's worth it, and negotiate a structure that doesn't quietly starve the company.

This is where the CFO stops being a scorekeeper and becomes a designer. You design policies that turn growth into cash instead of turning growth into a working-capital crisis. You design pricing guardrails so discounts don't accidentally sell product below its economic cost. You design approval thresholds so decisions get made at the right altitude. You design reporting that tells the truth early—not just what happened last month, but what the system is signaling about next quarter.

A CFO's strategic contribution often shows up in decisions that don't feel like "finance" decisions. For example: which customer segment you pursue, because the sales cycle and retention profile determine how much capital you need. Or: whether you hire senior talent now or later, because compensation structure changes burn rate and flexibility. Or: whether you build versus buy a capability, because capitalizing software development versus paying recurring SaaS fees affects both cash timing and operating leverage. The CFO doesn't have to own these calls, but you make them intelligible.

The co-pilot CFO also changes the nature of planning. Old planning is wishcasting: last year plus a growth rate, plus "strategic initiatives" that exist mostly as bullet points. Modern planning is about constraints and trade-offs. If you want to grow faster, what constraint breaks first: cash, people, systems, or customer success capacity? If you want higher margins, what must change: pricing, product mix, procurement, support model, or delivery efficiency? The CFO's job is to force the plan to answer those questions with numbers that can be executed.

One of the most valuable CFO moves is to replace vague optimism with staged commitments. Instead of approving a $5M project, you approve $500k to prove the riskiest assumption, then another tranche when the evidence shows up. This is capital allocation as discipline, not denial. It keeps the company ambitious without being reckless. It also changes internal politics: teams learn that finance is not a brick wall; it's a gateway that opens when proof arrives.

Stakeholders are the silent audience for all of this. Inside the company, the CFO serves operators who need clarity and fairness: managers want budgets they can trust, metrics that reflect reality, and variance explanations that lead to better decisions rather than blame. Outside the company, the CFO serves capital providers and gatekeepers: banks, bondholders, investors, auditors, rating agencies, regulators. They all have different languages, but they share one obsession: *can they trust you with other people's money?*

That trust is built in calm moments, not crisis moments. A CFO who only communicates when things are going well is like a pilot who only checks the weather when the sky is blue. The best CFOs run a steady cadence: not just reporting numbers, but explaining *drivers*. What changed in pipeline quality? What moved retention? Why did margin shift? What leading indicators are you watching? How does this quarter connect to the longer arc? Clarity reduces surprise; reduced surprise lowers the cost of capital.

There's a subtle but critical shift here: the CFO becomes a storyteller, but a different kind than marketing. This story is not about hype; it's about causality. You teach stakeholders how your business works so that when numbers wobble (and they always do), they interpret the wobble correctly. You earn the right to say, "We expected this dip because we pulled a lever on purpose," instead of sounding like you're making excuses after the fact.

And yes, sometimes the CFO still has to be the person who says, "No." But the mature version of "no" is not moral judgment. It's a statement about constraints. "No" might mean: we can't fund this without violating a covenant. Or: this pricing would create negative gross margin after support costs. Or: this acquisition would concentrate customer risk. Or: this hiring plan assumes productivity that our onboarding capacity can't deliver. A useful "no" comes with an alternative: "Here's what we can do instead, and here's what has to be true for the bigger move to become possible."

The CFO's relationship with the board is where the co-pilot role becomes visible. Boards don't want a CFO who is merely competent at reporting; they want an executive who can surface the real issues before they become board-level emergencies. The board is essentially buying insurance against self-deception. The CFO provides that insurance by making downside scenarios concrete and by proposing actions that preserve options: covenant headroom, liquidity runway, diversified funding sources, and early-warning indicators.

That doesn't mean you show up as the prophet of doom. Boards tune out constant alarms. The art is in distinguishing *risk* from *noise*. Risk is something that can kill you or permanently impair the business. Noise is volatility that looks scary but doesn't change the underlying engine. A CFO who learns that distinction becomes invaluable, because you help the organization avoid two expensive habits: panic and complacency.

The internal mandate is just as important: becoming the CFO of the whole company, not the CFO of the finance team. That means leaving the comfort zone of accounting and walking into the messy rooms: sales comp, product roadmaps, customer success capacity, inventory strategy, vendor negotiations. Not to micromanage, but to add a particular kind of thinking: *What does this do to cash? What does this do to unit economics? What does this do to risk concentration? What does this do to our ability to raise capital later?*

It also means building a finance function that can support this work. If your team spends three weeks closing the month and the fourth week explaining it, you have no time to shape decisions. Modern CFOs treat the close process like a manufacturing line: standardized, automated where possible, exception-driven, with clear ownership. The goal is not prettier spreadsheets; it's decision velocity. When numbers arrive quickly and reliably, the company can steer continuously instead of lurching.

Technology matters here, but not because "digital transformation" sounds impressive. It matters because finance is often the bottleneck for truth. When data is scattered, definitions are inconsistent, and reporting is manual, the organization starts debating whose spreadsheet is right instead of debating what to do. A CFO who invests in clean data models, consistent KPI definitions, and self-service dashboards is buying something rare: time. Time to analyze, time to partner, time to lead.

The CFO's influence also shows up in culture, whether you want it or not. The way finance behaves teaches the company what is safe to say. If finance punishes bad news, you'll get late bad news. If finance makes every budget conversation feel like a trial, you'll get sandbagging and gamesmanship. If finance treats forecasts as a tool for learning rather than a weapon for blame, you'll get earlier signals and better decisions. The CFO sets the tone for intellectual honesty.

One practical cultural lever is how you handle variance. Variance analysis can be a spotlight or a searchlight. A spotlight says, "Who messed up?" A searchlight says, "What changed in the system, and what do we do now?" The co-pilot CFO prefers searchlights. The goal is not to be nice; it's to be effective. Blame encourages concealment. Learning encourages disclosure. Disclosure is how you steer.

Another lever is incentives. Sales commissions, bonus plans, equity grants, and promotion criteria tell the company what it truly values. If incentives reward bookings without regard to margin or collectability, you'll get growth that looks great and cash that looks terrible. If incentives reward cost cutting without regard to customer experience, you'll get short-term margin and long-term churn. The CFO's job is to help leadership build incentives that align with durable value creation: growth that pays for itself, efficiency that doesn't hollow out the product, and investment that has a believable path to return.

Then there's the CFO's relationship with uncertainty. Many executives crave certainty because it feels professional. CFOs, ironically, are paid to live without it. Forecasts are guesses with math on them. Markets move. Competitors react. Customer behavior changes. The co-pilot CFO doesn't pretend the future is known; you build a system that remains stable even when the future is not. That means scenario planning, sensitivity analysis, and explicit risk limits—but done in plain language, tied to decisions people actually face.

For example: instead of presenting a single revenue forecast, you might present three scenarios and the operational commitments tied to each. In the downside scenario, we freeze hiring in non-critical roles and prioritize retention. In the base scenario, we hire in customer success and product. In the upside scenario, we accelerate sales hiring but only if churn remains below a threshold and sales cycle time holds. This isn't academic modeling. It's steering logic: "If the instruments read X, we do Y."

This is also where the CFO's personal operating style matters. The best CFOs are calm under pressure not because they're emotionally detached, but because they've rehearsed the decision tree. They know the levers: reduce discretionary spend, renegotiate vendor terms, slow hiring, adjust pricing, shift payment terms, raise capital, refinance, sell non-core assets, pause a project. Panic is often just unpracticed decision-making. The co-pilot CFO practices.

One more evolution: the CFO's mandate increasingly includes a broader view of value. Shareholders matter, obviously. But so do employees, customers, regulators, and communities—not as a moral afterthought, but because stakeholder trust affects operating freedom. When trust erodes, friction appears everywhere: hiring gets harder, customer acquisition costs rise, regulatory attention intensifies, and partners demand stricter terms. A CFO who understands value creation as a multi-stakeholder system isn't being political; you're being practical. The cheapest capital is credibility.

So what does "co-pilot" look like on an ordinary Tuesday, not in a dramatic board meeting? It looks like asking questions that sharpen decisions without hijacking them:

- "What assumption has to be true for this to work, and how quickly will we know if it's false?"
- "Which metric will move first if this goes wrong—cash, churn, margin, cycle time, or something else?"
- "If we're right, what does success look like in 30 days, 90 days, and 180 days?"
- "Are we buying growth, renting it, or financing it through our customers and suppliers?"

These questions are not "finance questions." They're leadership questions with financial consequences.

The CFO role, at its best, is a combination of spine and imagination. Spine: the willingness to insist on truth, controls, and solvency even when it's unpopular. Imagination: the ability to see paths to growth that are fundable, to structure bets intelligently, and to design an operating system that converts effort into value. If you only have spine, you become a gatekeeper. If you only have imagination, you become a gambler with a title. Co-pilot means having both, and knowing when each is required.

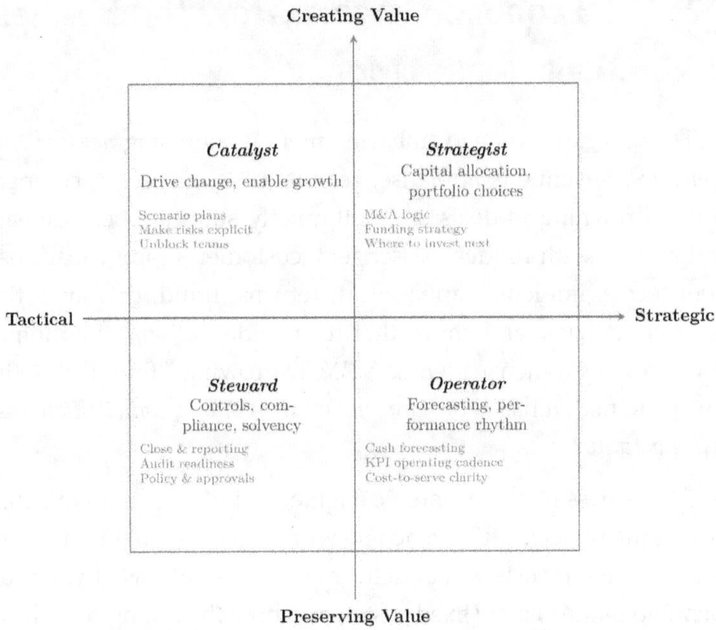

The most practical way to operationalize this role is to adopt a single, non-negotiable habit: before any material decision is approved, force it through a two-question gate. First: *What could make this decision fatal?* (liquidity, covenants, reputational damage, concentration risk, operational capacity). Second: *What would make this decision great?* (a measurable driver that improves cash, margin, retention, or strategic position). Write those answers down, attach an owner and a metric, and set an explicit checkpoint date. That simple discipline turns "finance" from a function into a steering system—and it turns the CFO from the keeper of rules into the keeper of momentum.

1.2 The Economic Engine: How the Business Actually Makes Money

A CFO's quickest route to embarrassment is confusing *revenue* with *economics*. Revenue is applause; economics is oxygen. A company can be drowning in deals and still quietly suffocate because each deal arrives with hidden passengers: customer support load, payment terms, implementation effort, returns, fraud, discounts that never come back, and churn that turns today's "win" into tomorrow's vacancy. When leaders say, "We're growing," the CFO's reflex can't be to nod. It has to be: *Growing what—volume, contribution, cash, or just activity?*

Every business has an economic engine, and it's less romantic than most strategy decks. It's a machine with a few essential parts: what you sell, what it costs you to deliver one more unit, what you must spend no matter what (fixed costs), and how the timing of cash collection compares to the timing of cash payments. If you understand those parts, you can predict a company's behavior under stress the way a mechanic can predict a car's behavior on a steep hill. If you don't, you end up in the classic executive argument: Sales celebrates top-line growth, Operations begs for headcount, and Finance wonders why the bank balance is dropping like a stone.

The phrase "business model" gets thrown around like it's a slogan—"We're a platform," "We're SaaS," "We're direct-to-consumer"—but those labels are mostly marketing. The CFO version of a business model is brutally simple: *how the company turns customer behavior into cash and profit*. Not what the product *is*. What the product *does* to the P&L and to the bank account. The same product can be a cash machine or a cash trap depending on pricing, delivery, and terms. Two companies can sell the same thing and live on different planets financially.

Start with the most basic question: *What is the unit?* Not the SKU in the system—the economic unit. For a grocery store, the unit might be a basket. For a software company, it might be a customer subscription or a seat. For a consulting firm, it's often a billable hour. For a marketplace, it's a transaction. Pick the unit that a real human on the other side actually chooses, because that's what determines demand. If you pick the wrong unit, everything downstream becomes fiction dressed up as precision.

Once the unit is clear, revenue becomes less mystical. Revenue is simply the price per unit times volume—with a caveat that CFOs learn the hard way: *the price you think you charge is rarely the price you actually collect*. Discounts, credits, refunds, chargebacks, rebates, and free months are not rounding errors; they are alternate realities. Revenue quality is part of the engine. It shows up in net revenue retention, returns, disputes, collections, and ultimately cash.

Then comes the most under-loved concept in corporate life: *variable cost*. Variable costs are the costs that rise when you sell more units: product materials, shipping, payment processing, usage-based infrastructure, commissions, fulfillment labor (if it scales with volume). These are the costs tied to the act of serving one more customer or delivering one more unit. The reason they matter is that they decide whether growth is nourishing or exhausting. If variable costs creep up as you scale—because you're expediting shipments, overstaffing support, or paying more to acquire customers who churn—your engine doesn't rev; it overheats.

The simplest, most useful metric that falls out of this is *contribution margin*. In plain language: it's what's left from a sale after you pay the costs that move with that sale. If you sell something for $100 and it costs $60 in variable costs to deliver, your contribution margin is $40. That $40 is the money available to pay for fixed costs and, after that, profit. This is why contribution margin is the CFO's favorite adult conversation. It doesn't care how inspiring your mission statement is. It asks one question: *Does the unit you're selling actually help pay for the company?*

A lot of internal debates disappear when contribution margin becomes the shared language. Marketing wants to spend more? Fine—but are we buying contribution or buying revenue? Sales wants to discount? Fine—but show how much contribution you're giving away and what volume increase is required to earn it back. Product wants to add a feature that raises hosting cost? Fine—but what does it do to contribution margin per customer, and what price or retention lift offsets it? Contribution margin turns opinions into solvable math.

Now meet the other half of the engine: *fixed costs*. These are the costs you pay even if you sell nothing tomorrow: leadership salaries, core engineering, rent, insurance, baseline systems, and often big chunks of payroll. Fixed costs are not "bad." They're a bet. When you choose fixed costs, you're saying, "We believe volume will be high enough that these costs become small per unit." That bet is what makes businesses like software magical at scale—and what makes them terrifying when growth stalls.

There's a particular drama that happens when companies confuse fixed and variable costs. A consulting firm can flex capacity by hiring contractors or using variable compensation; it can shrink when demand falls. A software company can't un-build the platform every time bookings dip. If it carries a large fixed cost base, it needs volume to justify it. That's not a moral failing; it's an operating choice. The CFO's job is to make sure everyone understands what kind of animal they're riding.

To make this concrete, imagine two companies with identical revenue: $10 million. One is a consulting firm with mostly variable delivery costs (people hired per project, subcontractors, travel billed back, commissions aligned with revenue). The other is a software company with a large engineering team and a product that costs little per incremental customer. The consulting firm's profit might be steady but capped: each dollar of revenue requires another slice of labor. The software company might lose money early but explode in profitability later: once fixed costs are covered, each additional customer contributes a lot of incremental profit. Same revenue. Completely different engine.

That difference has a name that shows up everywhere in finance: *operating leverage*. Operating leverage is a fancy term for a simple idea: *how sensitive profit is to changes in revenue*. High operating leverage means you have a lot of fixed costs relative to variable costs. When revenue rises, profit rises faster (after you cover fixed costs). When revenue falls, losses can deepen fast because the fixed costs don't politely shrink with demand. Low operating leverage means more of your costs move with volume, so profit changes more gently as revenue changes.

Operating leverage is the hidden reason some leadership teams feel like geniuses in a boom and feel cursed in a slowdown. The business didn't become smart and then dumb; the cost structure magnified the cycle. A CFO who can explain operating leverage without jargon does something powerful: you replace emotional whiplash with an understanding of the machine.

Here's the cleanest way to see it: *break-even*. Break-even is the point where contribution margin has finally paid for fixed costs. Before break-even, each unit sold reduces the loss but doesn't create profit. After break-even, each unit sold creates profit at roughly the contribution margin rate. That switch—from "selling reduces pain" to "selling creates profit"—is the moment high operating leverage becomes a jet engine.

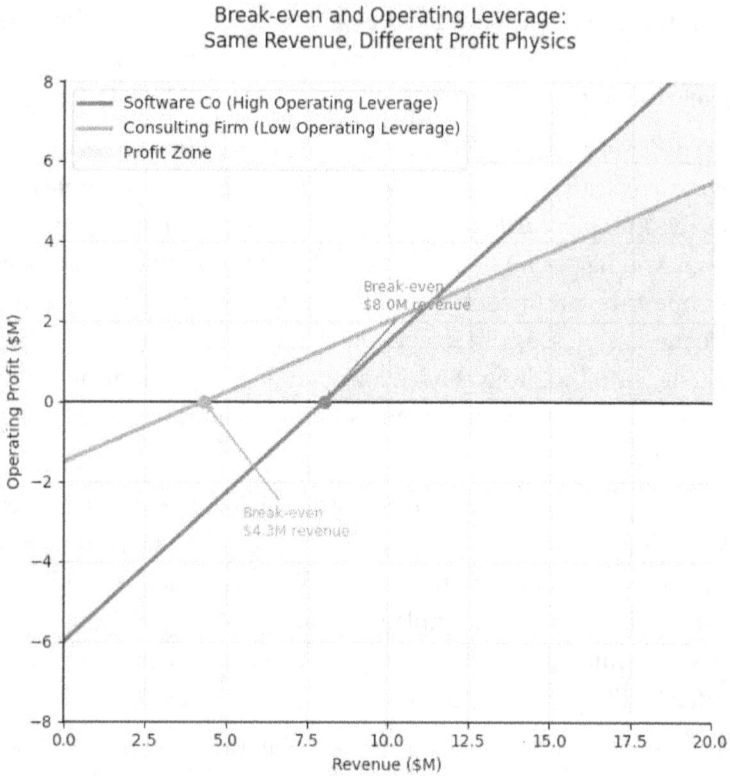

A break-even chart like this is more than a teaching tool; it's a negotiation tool. When a leader asks, "Why are we so obsessed with churn?" you can point to the slope. High operating leverage companies don't just lose revenue when customers leave; they lose the very thing paying for fixed costs. When a leader asks, "Why can't we just hire ahead of growth?" you can point to the break-even point moving to the right. Hiring ahead is not wrong, but it is a decision to increase fixed costs and therefore increase the revenue required to breathe comfortably.

Operating leverage is also why the same "efficiency initiative" can mean radically different things across business models. In a consulting firm, improving utilization (the percent of paid hours billed to clients) can be the difference between profit and loss, because labor is the product. In a software company, utilization is less meaningful; what matters might be gross margin on infrastructure, support cost-to-serve, and retention. A CFO who imports metrics from the wrong model creates confusion that looks like sophistication.

Gross margin deserves its own moment, because it's the most widely misunderstood number on the income statement. Gross margin is revenue minus cost of goods sold (COGS). In product businesses, COGS might be materials, manufacturing, shipping, and fulfillment. In software, COGS is often hosting, third-party data, support, and sometimes portions of implementation. Gross margin is not merely a percentage to brag about; it's the company's fuel efficiency. If your gross margin deteriorates as you grow, you're not scaling; you're simply driving faster while leaking gas.

In many modern companies, especially in tech and services hybrids, the real argument isn't about revenue growth. It's about what belongs in COGS versus operating expense. Teams can accidentally game this classification—not out of malice, but out of incentives and habit. The CFO's job is to enforce consistency so the business can learn. If support costs belong in delivering the product, put them in COGS so gross margin reflects reality. If you hide them below the line, you will overestimate how scalable the engine is and underinvest in fixing the true constraint.

Now zoom out to the full profit equation, still staying in human language:

$$Profit = (Revenue - Variable\ Costs) - Fixed\ Costs.$$

That first bracket is contribution. Contribution pays for fixed costs. What's left is profit. People will try to complicate this because complexity feels managerial. Resist. The power of the economic engine model is that it stays simple enough to drive decisions.

Once you have the engine model, you can ask the CFO questions that actually matter:

- If we discount by 10%, how much volume must increase to keep contribution dollars flat?
- If we add a layer of management, how much does break-even move?
- If we shift customer mix toward enterprise, what happens to sales cycle length, cost-to-serve, and cash timing?
- If we invest in automation, are we reducing variable costs (good) or just increasing fixed costs (possibly good, but a different bet)?

Notice what's missing: vibes. The engine doesn't care how confident the room feels.

The best CFOs develop a sensitivity to *where* the engine is fragile. High operating leverage businesses are fragile to demand shocks and churn. Low operating leverage businesses are fragile to labor market pressure and utilization drops. High gross margin businesses can still be fragile if customer acquisition costs are high and payback takes too long. Inventory-heavy businesses can be fragile to forecasting errors and obsolescence. Fragility is not shameful; it is simply the cost of a particular way of making money.

This is also where the CFO's "co-pilot" role becomes practical rather than philosophical. In leadership meetings, you're not there to recite finance vocabulary. You're there to connect operational choices to economic outcomes: "If we promise two-day delivery, here's what it does to shipping cost and return rates." "If we broaden the product line, here's what it does to inventory turns and working capital." "If we move upmarket, here's what it does to onboarding cost and the ramp time of the sales team." The engine makes cross-functional decisions speak the same language.

A simple table can help translate the engine into an operating map. The goal isn't to be exhaustive; it's to prevent category errors—using the wrong levers for the wrong engine.

Engine element	*What it really means*	*What executives often get wrong*
Unit	The customer choice that drives volume (seat, order, transaction, hour)	Tracking a proxy metric that looks measurable but doesn't drive cash
Price realized	What you actually collect after discounts, credits, churn, and disputes	Assuming "list price" is the economic truth
Variable cost	The cost of serving one more unit (COGS, fulfillment, usage, commissions)	Treating variable costs as "small" because they're spread across growth
Contribution margin	Dollars available to pay fixed costs and profit	Optimizing for revenue growth while contribution dollars stagnate
Fixed costs	The bet you place on future volume (team, systems, facilities)	Adding fixed cost because it feels like "investing," without moving break-even consciously
Operating leverage	How profit reacts to revenue changes (slope of the profit line)	Mistaking cyclical tailwinds for "a better business"

An engine model also acts like an honesty filter for popular metrics. Take LTV/CAC, the famous ratio comparing lifetime value (LTV) to customer acquisition cost (CAC). It's useful, but it's often built on sand. If LTV is calculated from gross margin while support, onboarding, and returns sit elsewhere, you overstate value. If churn is measured on a cohort that benefited from early discounts or unusual market conditions, you hallucinate durability. A CFO who anchors

LTV on *contribution* rather than on a flattering margin line is doing the organization a favor: you're preventing "profitable growth" from becoming a PowerPoint phrase that collapses in the bank account.

Even without fancy models, the CFO can upgrade the conversation by insisting on two numbers for any growth proposal: *incremental contribution dollars* and *incremental fixed cost dollars*. Incremental contribution is the fuel. Incremental fixed cost is the weight. The plan is viable if fuel grows faster than weight *in the timeframe you can afford*. That last clause matters, because survival is timed. A company can be a great business in three years and still miss payroll in six months.

Finally, remember that economic engines are not static. The most dangerous moment in a company's life is when the engine changes quietly. A service company starts building product and its fixed cost base rises. A product company adds custom implementation and its variable delivery costs rise. A marketplace subsidizes one side to grow and accidentally makes that subsidy permanent. The CFO's job is to call it what it is: "We are changing the engine." When you say it out loud, leadership can decide if it's intentional and fund it appropriately. When you don't, it happens anyway, and everyone acts surprised by the consequences.

Actionable takeaway: In the next leadership meeting where a major initiative is proposed (pricing change, hiring wave, new segment, product expansion), force a one-page "engine card" before approval: define the unit, state price realized assumptions, list variable costs per unit, compute contribution margin, list the fixed costs required, and show the new break-even revenue. If the team can't fill it out without hand-waving, the initiative isn't ready—not because it's bad, but because the company would be betting without knowing what game it's playing.

1.3 Financial Gravity: The Time Value of Money

The fastest way to lose control of a company's destiny is to treat time like a footnote. Executives do it constantly, usually without meaning to: "We'll make it up next quarter," "The savings show up over time," "The payback is quick." Time is the quiet variable hiding inside every sentence. It's also the variable that turns "profitable" into "bankrupt" and "great idea" into "bad deal." The time value of money is the name finance gives to this reality: a dollar now is not equal to a dollar later, because the calendar changes what that dollar can do.

If that sounds like a technicality, try this: would you rather receive $1,000 today or $1,000 a year from now? No tricks, no risk, guaranteed either way. Almost everyone chooses today. Not because people are impatient (though we are), but because money has *optionality*. With $1,000 today, you can pay down debt, invest, buy inventory, fund payroll, or simply keep it as insurance against a bad week. With $1,000 a year from now, you can do those things later—which is another way of saying you can't do them when they're needed most.

This is why time value of money (TVM) is less like a formula and more like gravity. Gravity doesn't care whether you're optimistic; it pulls anyway. TVM doesn't care whether you love the project; it discounts anyway. CFOs use it to answer the most adult question in business: *Is this future benefit worth today's cost?* If you can compare cash flows across time in a consistent way, you can compare projects, pricing structures, contracts, financing offers, and acquisitions without getting seduced by distant promises.

At a practical level, TVM gives you an exchange rate between *now* and *later*. That exchange rate is a rate of return (or discount rate)—a number that captures what money is worth in your world, given opportunity cost and risk. If your company can reliably earn 10% on incremental capital, then $1 today is "worth" more than $1 next year because next year's dollar is missing the chance to become $1.10. If capital is scarce, the exchange rate gets harsher; time becomes expensive.

Two operations sit at the heart of TVM, like gears that spin in opposite directions:

- *Compounding*: pushing today's money into the future.
- *Discounting*: pulling future money back to today.

Compounding is the part people intuitively like. Discounting is the part people argue with. The CFO's job is to make both feel obvious, because the business will otherwise behave like a person who only believes in calories when eating cake, not when stepping on a scale.

Compounding is straightforward: money grows over time when it earns a return. If you have a present value (PV) of $100 and it earns 10% per year, after one year you have a future value (FV) of:

$$FV = PV(1 + r)^t = 100(1.10)^1 = 110.$$

After two years:

$$FV = 100(1.10)^2 = 121.$$

The exponent is the point. Time is not additive; it's multiplicative. Returns earn returns. That's why long horizons make small differences in rates explode into huge differences in outcomes.

Discounting is the mirror image: future money is worth less today because you could have invested today's money and ended up with more in the future. So you divide instead of multiply:

$$PV = \frac{FV}{(1 + r)^t}.$$

If someone promises you $121 in two years and the relevant rate is 10%, the present value is:

$$PV = \frac{121}{(1.10)^2} = 100.$$

The math is clean, but the behavioral implication is spicy: the further out a cash flow is, the less it matters *today*. That doesn't mean long-term thinking is wrong. It means long-term thinking must be paired with realistic rates and credible paths to cash.

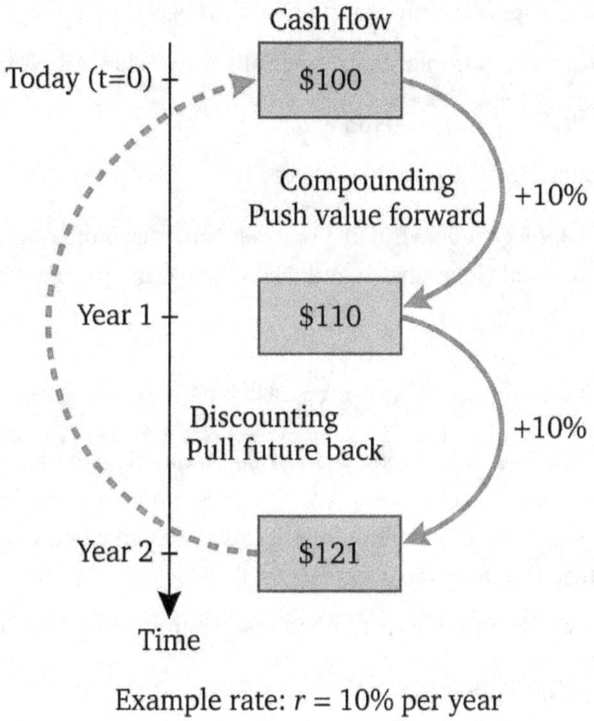

That picture is the CFO's daily mental model. It's also the antidote to one of the most common corporate illusions: treating a payoff that arrives "eventually" as if it were already in hand. "We'll be profitable once we scale" may be true. The question TVM forces is: *How much cash do we burn before we reach scale, and what is that burn worth in today's dollars?* Time doesn't just discount benefits; it magnifies costs when those costs are funded with scarce capital.

Discounting becomes even more important when you're comparing options that are intentionally structured to confuse you. Vendors love multi-year contracts with discounts, customers love payment terms that stretch, and acquisitions love earn-outs that postpone the painful part of the price. None of these are automatically bad. They're just time games. A CFO who can discount cash flows can turn time games back into comparable numbers.

Take a common example: a customer offers two choices.

- Option A: Pay $1,000 today.
- Option B: Pay $1,100 in one year.

Option B looks bigger. But if your relevant discount rate is 15% (because capital is expensive and risk is real), the present value of $1,100 in one year is:

$$PV = \frac{1100}{1.15} \approx 956.52.$$

So Option B is actually *worse* than $1,000 today. The extra $100 is a mirage created by time. This is why CFOs often push for upfront payments or milestone billing: not because they like being difficult, but because the calendar is a cost.

Now invert the situation: you're the one paying. A supplier offers:

- Pay $1,000 today, or
- Pay $1,100 in one year.

If you can earn more than 10% on cash (or avoid borrowing at more than 10%), deferring payment may be rational. TVM doesn't tell you what to do morally; it tells you what you're actually choosing economically. It replaces vague instincts with a price tag on timing.

Once discounting clicks, the leap to project evaluation is natural. Most strategic investments have the same shape: cash outflows early (build, hire, market, implement), then inflows later (revenue, savings, margin improvement). Your job is to add those inflows and outflows on the same footing. You do that by converting every future cash flow into a present value and summing them. That sum is called net present value (NPV):

$$NPV = \sum_{t=0}^{T} \frac{CF_t}{(1+r)^t}.$$

If NPV is positive, the project creates value relative to the discount rate. If it's negative, it destroys value. Again: not a moral judgment, but an economic one.

The sneaky part is that NPV is sensitive to assumptions that teams love to bury. When someone says, "The payback is two years," they're often hiding what happens *after* two years and what had to be true to get there. When someone says, "The terminal value is huge," they're often hiding how fragile far-off cash flows are under discounting. CFOs earn their keep by dragging those assumptions into daylight and insisting on ranges, not just point estimates.

TVM also forces you to confront risk in a disciplined way. A dollar five years from now is not only delayed; it is uncertain. In finance, we often capture that uncertainty in the discount rate: higher risk, higher rate, lower present value. This is why a risky growth story can sound thrilling and still be worth surprisingly little in present value terms if the cash flows are far away and the rate is high. That's not pessimism; it's a reminder that risk and time compound in the *wrong* direction for valuation.

This is where CFO judgment matters. Discount rates are not plucked from the air, but they are also not handed down on stone tablets. In practice, companies use hurdle rates (minimum acceptable returns) that reflect their cost of capital, risk appetite, and the scarcity of funding. A cash-rich utility and a venture-backed startup should not use the same discount rate. They are living under different gravity.

Consistency is another CFO landmine. You must match the cash flows to the rate. If cash flows are *nominal* (they include inflation), the discount rate should be nominal. If cash flows are *real* (inflation removed), the discount rate should be real. Mix them and you'll get an answer that looks mathematical and is mechanically wrong. The same goes for timing conventions: annual cash flows discounted annually, monthly cash flows discounted monthly, or at least adjusted appropriately. TVM rewards careful thinking and punishes casual mismatches.

Compounding conventions create their own subtle traps. Banks and lenders love quoting APR (annual percentage rate) while compounding monthly or daily. The effective annual rate (EAR) is what you actually pay or earn after compounding. If an APR of 12% compounds monthly, the effective annual rate is:

$$EAR = \left(1 + \frac{0.12}{12}\right)^{12} - 1 \approx 12.68\%.$$

That difference is not academic when you're pricing financing, evaluating customer payment terms, or comparing investment options. The CFO doesn't need to perform this calculation in a meeting from memory; you need the instinct to ask, "Is that APR or effective? What's the compounding frequency?"

Here's an uncomfortable truth about most "strategic" projects: the real decision isn't whether the project works in a spreadsheet. It's whether the company can *survive* the time gap between spending and earning. TVM is the language of that gap. It lets you translate the gap into a value and a runway requirement. That's why CFOs care so much about cash flow timing, milestone billing, implementation calendars, and ramp curves. If the timing is wrong, a good project becomes a liquidity event—and not the fun kind.

This is also why TVM should show up in pricing decisions, not just capital budgeting. Consider two pricing models for the same product:

- $0 upfront and $500 per month for 36 months, or
- $12,000 upfront and $250 per month for 36 months.

Which is better? Sales will argue about deal friction, Product will argue about customer psychology, and everyone will pick the one that feels aligned with strategy. The CFO should run both through a present value lens using a rate that reflects your funding reality. The "best" option might not be the one with the higher total dollars; it might be the one that pays earlier, reducing financing needs and increasing optionality.

TVM has a sharp edge in negotiations, too. When a counterparty offers an earn-out, they're often saying, "Let's postpone payment until later, and we'll make it contingent." That might align incentives. It might also reduce the deal's true value after discounting. If a seller insists on a higher headline price but most of it is deferred, discounting can reveal the truth: the headline is for press releases; the present value is for decision-making. CFOs who can calmly explain that distinction change the tone of negotiations from ego to economics.

One more place TVM quietly rules: performance targets. Many incentive plans reward revenue or EBITDA in the current year, which can encourage behavior that pulls value from the future into the present (over-discounting, underinvesting, cutting preventative maintenance, delaying necessary hires). TVM reminds you that value is a stream across time, not a single snapshot. A CFO who brings present value thinking into incentive design helps prevent the organization from cannibalizing tomorrow to decorate today.

None of this requires turning everyone into quants. The goal is to build a shared instinct: *time has a price*. When teams internalize that, they naturally start asking better questions:

- "How soon do we get paid?"
- "What's the cash payback, not just the accounting payback?"
- "If the benefit arrives in year 5, what has to be true in years 1–4 to survive?"
- "Are we being compensated for the risk of waiting?"

Those are CFO-grade questions, and they're infectious in the best way.

Actionable takeaway: For the next major decision involving delayed benefits (a long-term contract, a multi-year project, a pricing change with deferred cash, or an acquisition earn-out), require a one-page present value view: list the expected cash flows by period, pick a single discount rate with a written rationale, compute the present value of each cash flow, and highlight which year contributes most of the present value. If most of the "value" lives far in the future, don't kill the idea—tighten the structure (milestones, upfront payments, staged investment) until the calendar stops being your largest hidden risk.

Chapter 2

The Scoreboard: Financial Statements and Reporting

Imagine coaching a professional sports team where the scoreboard is hidden during the game, or worse, the score is updated three weeks after the match ends using random calculations. That is what running a business feels like without solid financial reporting. In this chapter, we move beyond the dry mechanics of debits and credits to build your ultimate dashboard. We will explore how to translate chaotic daily transactions into a clean narrative of performance, ensuring you aren't just counting beans, but actually keeping score to win.

2.1 Accrual Accounting and Policies

A customer buys your annual plan on January 2. Your bank balance jumps. Slack celebrates. Someone says, "We're profitable now." Meanwhile, your team still has to deliver twelve months of service, pay the cloud bill every month, and answer support tickets on a random Tuesday in October when the customer forgets their password. That single moment—cash arriving early while work and costs arrive later—is the quiet engine behind most financial confusion. Cash feels like reality because it's tangible. But cash is also *bad at telling time.* Businesses don't live in single moments; they live in stretches of obligation and delivery. Accrual accounting exists because decision-makers need a clock, not just a bank statement.

Cash accounting tells a story with only one verb: *paid*. Accrual accounting tells a story with two verbs: *earned* and *incurred*. "Earned" means you delivered what you promised (or enough of it) and can reasonably claim the revenue. "Incurred" means you consumed resources to run the business—whether you've paid for them yet or not. The point is not to be fancy. The point is to stop confusing timing with performance. A company can look flush because it collected a pile of annual prepayments and still be running a money-losing machine underneath. A different company can look broke because customers pay late, while its underlying unit economics are strong. Cash measures *liquidity*. Accrual measures *economics*. You need both, but you can't ask one to do the other's job.

A useful way to feel the difference is to picture a credit card swipe. You walk out of the store with the headphones today; the cash leaves your account later. Your life doesn't treat those as the same moment. You mentally record the purchase when you get the headphones, not when the bill is paid. Businesses have the same problem, just with more zeros and more ways to hide the truth. Accrual accounting is the business version of admitting, "We already committed to this," or "We already benefited from this," even if cash hasn't moved yet.

The core idea is the matching principle: expenses should be recognized in the same period as the revenue they helped generate. It sounds polite—match, align, be tidy—but it's really a weapon against self-deception. If you sell a twelve-month subscription today and book all the revenue today, you'll look like a hero in January and a disappointment from February onward. If you pay an annual insurance policy up front and book all the expense in one month, you'll look incompetent in that month and oddly brilliant for the next eleven. Matching smooths the story so the scoreboard reflects what happened *economically* across time, not what happened to the checking account on a particular day.

Here's the subscription example in plain mechanics. When the customer pays you $12,000 for a one-year plan on January 2, you did not "earn" $12,000 that day. You received $12,000 and you gained an obligation: deliver service for the next twelve months. Accrual accounting records that obligation as *deferred revenue* (a liability). Each month, as you deliver service, you recognize $1,000 of revenue and reduce the liability by $1,000. The cash was a lump. The economics are a stream.

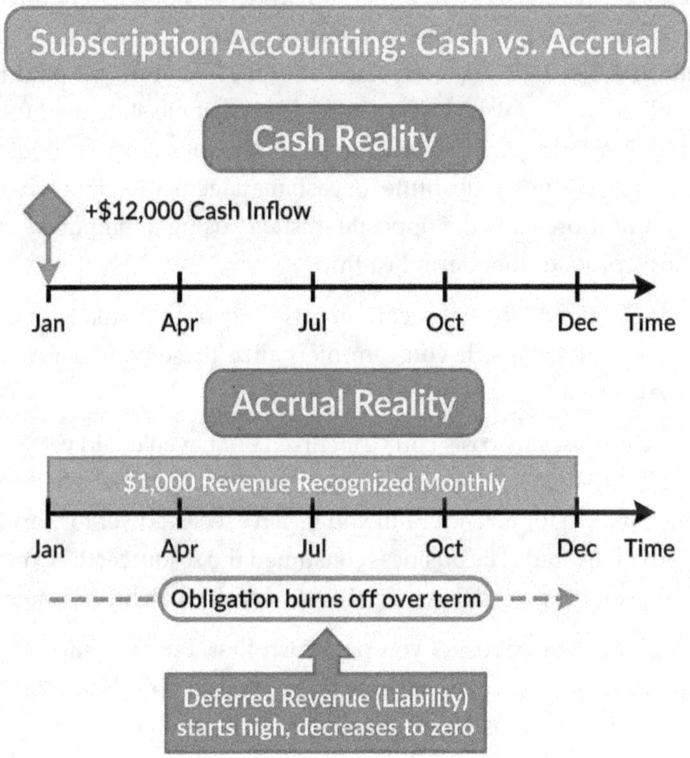

The matching principle isn't only about aesthetics; it changes what you can safely conclude. Imagine you're deciding whether to hire two more engineers in February. Cash accounting might show a huge January "profit" because the annual invoice landed, suggesting you can afford the hires. Accrual accounting will show the more sobering reality: you earned only one-twelfth of that sale in January, and you'll need to keep earning the remaining eleven-twelfths by continuing to deliver. The hiring decision becomes tied to the durability of your revenue stream, not the luck of when an invoice got paid.

Accrual accounting also explains why some of the most dangerous businesses look healthy right up until they don't. A company can extend generous payment terms to win deals, book revenue when the product ships, and show impressive growth on the income statement. If collections lag, cash starts starving. Vendors get paid late. Payroll becomes a weekly horror film. The income statement might still look strong while the business quietly approaches a liquidity cliff. Accrual is not a substitute for cash management—it's a separate lens—but it prevents the opposite mistake: using a temporary cash spike as proof of long-term health.

Under the hood, accrual accounting is built from a small set of recurring adjustments. If you can internalize these, you can read financials like a native.

Accrued expenses are costs you've incurred but haven't paid yet. Think payroll earned in the last week of the month but paid in the first week of next month, or a vendor bill you haven't received yet for services already delivered. The business consumed the resource; the expense belongs to this period, even if accounts payable catches up later.

Prepaids are the opposite: you paid cash first, but you haven't consumed the benefit yet. Insurance is the classic. You didn't "spend" twelve months of insurance in one day; you purchased the right to be covered over time. A prepaid asset gets expensed gradually as time passes.

Deferred revenue is cash received before you've earned it. Subscriptions, retainers, implementation fees tied to milestones—anything where the customer paid for future delivery. Deferred revenue is often mistakenly celebrated as "free money." It isn't. It's a promise with a due date.

Non-cash expenses are where accrual accounting stops being intuitive and starts being powerful. Depreciation and amortization spread the cost of long-lived assets across the periods that benefit from them. You pay cash for equipment once, but you "use up" the equipment over years. The expense is the wearing out of the asset, not the purchase transaction. This is one reason net income and cash flow diverge: depreciation reduces profit without reducing cash in that moment. That divergence is not a trick; it's the accounting system refusing to lie about how resources are consumed.

All of those adjustments are driven by policies. "Policy" can sound like bureaucracy—something written to please auditors and punish joy. In practice, accounting policies are the rules of the game. They determine what "revenue" means in your business model, what counts as an "expense" this month versus next month, and how much judgment is allowed in the gray areas. Two companies can sell the same product, collect the same cash, and report different profits because their policies answer timing questions differently.

A good policy does two jobs at once: it reflects economic reality *and* it creates consistency. Consistency is underrated until you've lived through its absence. Without it, the monthly P&L becomes a mood ring—green when someone feels optimistic and red when someone decides to be "conservative." Investors, boards, and operators can handle bad news. They cannot handle numbers that change meaning every month.

Revenue recognition is the most emotionally charged policy because it decides when you're allowed to call success "success." In many businesses, revenue is not a single event; it's a sequence: contract signed, implementation delivered, product shipped, service consumed, renewal earned. A crisp policy answers: at what point have we actually delivered enough value to earn revenue? The answer depends on what you promised and how you deliver it. The finance team's job is to translate the promise into a repeatable rule that doesn't flinch under pressure.

Expense capitalization is another policy that invites creative storytelling. Some costs are plainly period expenses: rent, utilities, most salaries. Others can be capitalized—recorded as an asset—when they create future benefit beyond the current period. Software development, major implementation projects, and certain acquisition-related costs can land in this territory depending on the accounting framework and facts. Capitalizing costs increases current-period profit because you're not expensing them immediately. It also creates future expense (amortization) as the asset is written down. A thoughtful policy prevents the organization from turning capitalization into a profit lever that managers pull when they want a prettier month.

Then there are estimates: the built-in humility of accrual accounting. Accrual doesn't only shift timing; it forces you to guess—carefully, transparently, and consistently. Allowance for doubtful accounts is an estimate of how much receivables you won't collect. Warranty reserves estimate future costs tied to today's sales. Returns reserves estimate how much of today's revenue will come back. Useful lives for depreciation estimate how quickly assets wear out. These estimates are not weaknesses; they're an admission that financial statements are a model of reality, not reality itself. The danger is not estimating. The danger is estimating without discipline.

Discipline comes from documentation and governance. When a company says "our revenue is $X," sophisticated readers want to know: according to what definitions? Under what terms? With what exceptions? This is why the notes and the "significant accounting policies" matter. They are the legend of the map. Without the legend, the numbers are just colored shapes.

The most practical expression of policy is the chart of accounts (CoA). People treat the CoA like a dusty spreadsheet full of numbers that only accounting cares about. That's like calling the skeleton "a bunch of bones." The chart of accounts is the structure that determines whether your data becomes insight or sludge. It's the filing cabinet, the labeling system, and the lens through which management sees the business.

A chart of accounts is a list of every bucket you can legally and operationally put money into: revenue accounts, expense accounts, asset accounts, liability accounts, and equity accounts. But the real design question is not "how many accounts?" It's "what decisions should this set of accounts make easier?" If your CEO asks, "Are we making money on this product line?" and your CoA can't separate that product's direct costs from everything else, the question becomes a research project instead of a dashboard glance. If your head of sales asks, "Did discounting increase churn?" and you don't track discounts or separate customer success costs, you will argue from anecdotes.

CoA design lives in the tension between clarity and chaos. Too few accounts and everything becomes a miscellaneous soup. Too many and the organization drowns in coding errors, inconsistent usage, and reports no one trusts. The sweet spot is a CoA that captures the major economic drivers of the business and aligns with how leaders actually operate: product lines, customer segments, geographies, channels, and key cost categories. Often, the best answer isn't an explosion of accounts—it's a combination of a reasonably sized CoA plus dimensions (classes, departments, locations, projects) that allow slicing and dicing without multiplying accounts into oblivion.

Think of the CoA as the difference between a pantry and a junk drawer. A pantry has categories that reflect real usage: grains, canned goods, spices. You can find what you need quickly, you notice when you're low on something important, and you don't buy duplicates because you forgot what you already had. A junk drawer technically "stores" items too, but it doesn't support decisions—only scavenger hunts. Many companies run their finances out of a junk drawer and then wonder why forecasting feels like astrology.

A policy-driven CoA also prevents a subtle but common failure: mixing operating and non-operating items until performance becomes unreadable. If one-time legal settlements, restructuring charges, or unusual consulting projects land in the same accounts as normal operating spend, monthly trends will whipsaw and managers will lose trust. The goal isn't to "hide" ugly items; it's to label them honestly so operating performance can be evaluated without constantly asking, "Is this real, or is this weird?"

Accrual accounting becomes especially valuable when you pair it with operational reality. Working capital is where the two worlds meet: receivables, payables, and inventory. These balance sheet accounts are often treated as accounting artifacts, but they are operational behaviors frozen into numbers. Receivables reflect billing and collections discipline. Payables reflect vendor terms and payment cadence. Inventory reflects purchasing, production planning, and demand accuracy. Accrual accounting captures these as assets and liabilities, then the cash flow statement later tells you whether those balances expanded or contracted. If you understand the accrual logic, you can look at a spike in receivables and ask the right operational question: "Did we sell more, bill later, collect slower, or dispute more?" Those are very different problems, and the ledger can help you diagnose which one you're facing.

There's also a human side to policies: they protect teams from politics. Without written, agreed rules, every close becomes a negotiation. Sales wants revenue now. Operations wants costs pushed out. Leadership wants the month to look "smooth." Accounting becomes the referee in a game where the rules are invented mid-play. Clear policies turn those conversations into something healthier: "Here's how we treat it every time." Consistency doesn't remove judgment, but it limits the places where persuasion can masquerade as accounting.

That said, rigidity can be its own form of dishonesty. The best finance leaders write policies that are specific enough to be repeatable, but flexible enough to reflect genuine changes in the business. When your revenue model changes—say you move from one-time licenses to subscriptions—your policies must evolve. When you launch a new product with materially different delivery obligations, your revenue recognition approach may need a new playbook. Policy is not meant to fossilize the business; it's meant to keep the scoreboard legible while the game changes.

A practical way to pressure-test your accruals and policies is to run a simple monthly thought experiment: "If we froze sales today, what would still be true next month?" You would still owe payroll. You would still deliver service to prepaid customers. You would still collect some receivables. You would still pay down some payables. That exercise forces you to see the business as a set of ongoing obligations and entitlements, not just a stream of cash movements. The balance sheet becomes less mysterious when you think of it as the list of things that remain true even when activity stops.

Here's another test, even more CFO-ish: take any large P&L line and ask, "What is the balance sheet partner to this?" Revenue partners with receivables and deferred revenue. Payroll expense partners with accrued payroll. Rent expense partners with prepaid rent and lease liabilities (depending on your lease accounting). When a P&L line has no plausible balance sheet partner, it may be fine (some expenses are simply paid and consumed in the same period), but it's worth questioning. It's a fast way to find missing accruals and sloppy categorization.

The irony is that accrual accounting, when done well, feels less like accounting and more like storytelling with guardrails. It insists that you acknowledge future obligations created by today's cash. It insists that you record the cost of resources when you actually consume them. It insists that you label things consistently so trends mean something. It doesn't guarantee good outcomes, but it dramatically improves the odds that you will *notice* reality early enough to respond.

Actionable takeaway: Pick one recurring transaction in your business that creates timing confusion—subscriptions collected up front, annual insurance paid in advance, payroll timing, large implementation projects. Write a one-page policy for it that answers three questions:

- When is revenue (or expense) recognized, and why?
- What balance sheet account carries the timing difference (deferred revenue, prepaid, accrual, etc.)?
- What is the monthly journal entry pattern?

Then audit the last three months and quantify the difference between cash timing and accrual timing for that one item. If that single policy makes your monthly results calmer, more comparable, and easier to explain, you've just bought back decision-quality with almost no new software—only clearer thinking.

2.2 The Three Key Statements

A finance leader's superpower isn't spreadsheet speed—it's being able to look at three documents and immediately sense when the story doesn't add up. The P&L says you "made" $800k, the bank account somehow shrank, and the balance sheet shows receivables ballooning like a parade float. None of those facts are inherently alarming. What's alarming is not knowing which one is lying, which one is early, and which one is simply answering a different question.

The trick is to stop treating the financial statements like three reports and start treating them like one machine viewed from three angles. The income statement (P&L) is performance over time. The balance sheet is your position at a point in time. The cash flow statement is the bridge that explains how you moved from one cash balance to another. When they're built correctly, they *must* reconcile. If they don't, the issue is never "accounting is weird." The issue is that something is missing, misclassified, or misunderstood.

The P&L is where the ego goes to either swell or get bruised. Revenue is the headline number everyone repeats, but the P&L's real job is more surgical: it answers whether your operations created economic value during the period, using accrual rules to place revenue and expense in the right months. That's why a well-designed P&L feels like a movie-motion, causality, momentum. You see whether gross margin is improving, whether operating expenses are scaling, and whether the business model is gaining leverage or just gaining complexity.

There's also a practical reason leaders obsess over the P&L: it's the most easily gamed statement without technically committing fraud. A month can be "saved" by pulling forward revenue recognition, pushing out expenses, capitalizing costs that probably shouldn't be capitalized, or classifying recurring spend as "one-time." The P&L is the statement most exposed to human incentives. That doesn't make it untrustworthy; it makes it a place where a CFO needs policies, consistency, and a low tolerance for creative writing.

If the P&L is the movie, the balance sheet is the freeze-frame—and it's where the plot twist usually lives. The balance sheet is not a dumping ground for "stuff that isn't on the P&L." It is a disciplined list of what the business *has* (assets), what it *owes* (liabilities), and the residual claim after those two fight it out (equity). It's the scorecard of accumulated decisions: every credit term you offered, every vendor you paid late, every inventory bet you made, every software build you capitalized, every dollar of profit you kept instead of distributing.

Two balance sheet lines explain an uncomfortable amount of corporate drama: accounts receivable and deferred revenue. Receivables say, "We recorded revenue we haven't collected yet." Deferred revenue says, "We collected cash we haven't earned yet." One is optimism awaiting proof. The other is a promise awaiting delivery. Both can be healthy. Both can also be early warning systems. A swelling receivables balance can mean strong sales—or weak collections—or disputed invoices—or customers quietly choking. A shrinking deferred revenue balance can mean you're recognizing revenue faster than you're selling new contracts, which is a polite way of saying the future might be thinner than the present.

Then there's the cash flow statement, the document that ends most arguments. The cash flow statement doesn't care about how hard you worked, how good the quarter "felt," or how many invoices you sent. It tracks what happened to cash, classified into operating, investing, and financing activities. That classification matters because cash can rise for opposite reasons. Operating cash rising is usually a sign of a business that funds itself. Financing cash rising can mean you took on debt or sold equity to keep going. Investing cash falling can mean you're building capacity for the future—or buying toys you can't afford. Cash is honest, but it's not judgmental. It will happily reflect both genius and denial.

Most operating cash flow statements you'll see use the indirect method, which starts with net income and then reverses and adjusts until you're back to actual cash from operations. This is where accrual accounting gets audited by reality. Depreciation gets added back (non-cash). Changes in working capital accounts show up as sources or uses of cash. If receivables increased, you subtract the increase (you booked revenue without collecting). If payables increased, you add the increase (you incurred expense without paying). It's not magic; it's algebra applied to the balance sheet.

Here's the uncomfortable but liberating truth: the cash flow statement is mostly the P&L and two balance sheets wearing a trench coat. Once you see that, you stop memorizing formats and start checking linkages.

Statement Linkages: the math that makes the story non-negotiable

The most important line in the P&L is net income—not because it's sacred, but because it has obligations. Net income doesn't get to just sit there looking impressive. It has to go somewhere. In accounting, that "somewhere" is equity, specifically retained earnings (profit kept in the business rather than paid out). If a company is profitable and pays no dividends, equity should increase over time. If it doesn't, you don't have an accounting mystery; you have a broken linkage.

Cash has a similar obligation. The ending cash balance on the cash flow statement must match the cash line on the balance sheet. Always. If it doesn't, you're not debating interpretation—you're debugging. This single linkage is why seasoned finance teams treat cash as the ultimate reconciliation anchor. Cash is where errors get trapped because cash is the one number that can be independently confirmed with a bank statement.

Those two anchors—net income flowing into equity and ending cash matching the balance sheet—create a loop. Inside that loop are the bridge mechanics: non-cash items and changes in balance sheet accounts that explain why net income and cash from operations diverge. When people say, "Our profits are up but we're burning cash," they're usually describing one of three things: receivables growing (customers haven't paid), inventory growing (you bought future sales), or deferred revenue shrinking (you're delivering past sales faster than you're selling new ones). The statements won't just tell you that those things happened; they'll quantify them.

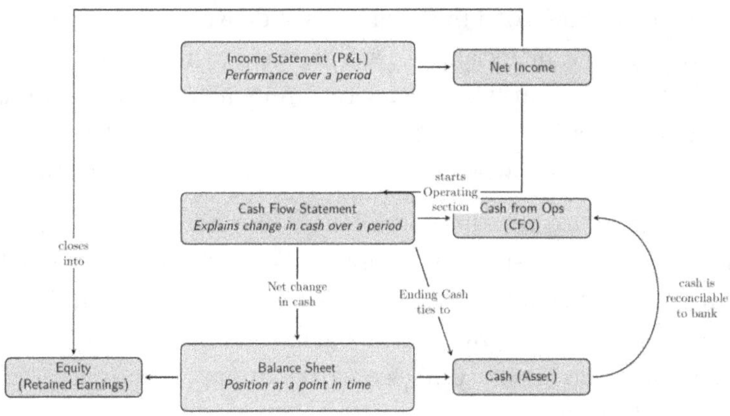

Once those linkages are in your bones, you can do something that looks like intuition but is actually practice: you can infer operational behavior from accounting movement.

If net income is strong but operating cash flow is weak, look for working capital expansion. A growing business can be "profitable" and still starve because it's funding customers through receivables, funding production through inventory, or paying down payables faster than it should. In other words, you can be winning on paper while financing the win with your own bloodstream.

If operating cash flow is strong but net income is weak, you might be seeing the opposite: deferred revenue growth (cash collected early), payables stretch (cash conserved), or non-cash expenses (depreciation) depressing earnings. This can be healthy or it can be a sign you're borrowing time. Deferred revenue is a wonderful cash engine—until churn rises. Payables stretch is a useful lever—until vendors tighten terms or service quality suffers. The statements don't moralize; they tell you which lever you pulled.

How to read each statement like a CFO (without turning into a forensic accountant)

On the P&L, start with gross margin, not revenue. Revenue is often a measure of how good sales is at signing deals. Gross margin is a measure of how good the business is at delivering value without destroying itself financially. A SaaS company with 85% gross margin has room for mistakes; a services firm with 35% gross margin has to be disciplined like a pilot. If gross margin is deteriorating, everything else is downstream pain, even if revenue is climbing.

Next, scan for expense lines that behave like mood swings: professional services that spike randomly, marketing that comes in bursts without a clear narrative, "other" that grows like a weed. These aren't just accounting hygiene problems; they're management visibility problems. If you can't see what's driving spend, you can't manage it—only react to it.

On the balance sheet, the fastest CFO-level read is a working capital gut check. Compare receivables growth to revenue growth. Compare inventory growth to cost of goods sold growth. Compare payables growth to cost of goods sold growth. You're not doing a full model in your head; you're looking for mismatches that suggest either operational friction or accounting timing issues.

Also watch the "soft asset" zones: capitalized software, deferred contract costs, goodwill, and other intangibles. These aren't automatically bad, but they're areas where optimism can be recorded as an asset. If those balances rise quickly, ask what economic benefit is expected and how it will be tested over time. The balance sheet is where hope gets capitalized.

On the cash flow statement, separate operating cash from financing cash with almost religious insistence. If cash is rising because you borrowed money or raised equity, that's not operational strength; it's fuel. Fuel is great—until you confuse it with an engine. The question isn't whether you used fuel. The question is whether the engine is getting more efficient as you scale.

One more subtlety: cash flow statements usually exclude non-cash investing and financing transactions, even though they can be economically huge. Leasing equipment, converting debt to equity, issuing stock for an acquisition—these may not move cash today, but they absolutely change obligations and ownership. Good reporting discloses them, and a CFO reads those disclosures the way a pilot reads weather.

A concrete mini-example: why profit and cash can diverge without anyone doing anything ``wrong"

Assume a company sells $1,000,000 of product in a quarter, all on 60-day terms. It recognizes $1,000,000 of revenue on the P&L. Cost of goods sold is $400,000, paid immediately to suppliers. Operating expenses are $450,000, mostly payroll paid in the quarter. Net income is $150,000.

Now assume customers are slow to pay, and only $600,000 of the $1,000,000 is collected by quarter-end. Cash impact:

- Cash in from customers: $600,000
- Cash out for COGS: $400,000
- Cash out for OpEx: $450,000

Net cash change from operations is $-\$250,000$, even though net income is $+\$150,000$. The difference is the $400,000 increase in accounts receivable (revenue recognized but not yet collected). Nobody cheated. The business simply financed customers, and the cash flow statement tattled.

This is why the statement linkages matter operationally. "We're profitable but cash is tight" isn't a complaint; it's a diagnosis prompt. Tight cash with profits usually means the business is growing, investing, or being generous with terms. The right response could be tightening collections, changing billing cadence, renegotiating payment terms, or raising working capital financing. The wrong response is pretending the P&L will pay payroll.

A CFO's checklist for statement sanity

A clean set of statements produces a few invariants—quick checks that catch a shocking number of issues before they become boardroom embarrassment.

If net income is positive for the year and there were no dividends or major equity transactions, retained earnings should rise. If it doesn't, find out what hit equity directly (prior period adjustments, stock compensation entries, currency translation, reclassifications). The "why" can be legitimate, but it should never be mysterious.

Ending cash on the cash flow statement must equal cash on the balance sheet. If your team can't tie those two within minutes, stop the meeting and fix the process. This is the accounting equivalent of checking that the brakes work before driving down a mountain.

The change in each balance sheet account should have an explanation that sounds like the business. "Receivables increased because collections slowed due to a billing system change" is a business explanation. "Receivables increased because... accounting" is not. The numbers are supposed to be the business, expressed in dollars.

Actionable takeaway: Take your most recent monthly package and do a 20-minute linkage drill. (1) Tie ending cash on the cash flow statement to cash on the balance sheet. (2) Tie net income to the change in retained earnings, adjusting for dividends and equity activity. (3) Pick the single largest working-capital change (AR, inventory, AP, deferred revenue) and write one sentence explaining it in operational language. If you can't do all three cleanly, the issue isn't your business model—it's the plumbing that turns transactions into truth.

2.3 The Close Process

On the last day of the month, reality is messy. Someone forgot to submit an expense report. A vendor invoice is stuck in an approval chain. A customer swears they paid (they didn't), and Sales is lobbying hard to "just book it" because the contract is signed. Meanwhile the CEO wants numbers by Monday, a department head wants to know if they can hire, and the board deck is due. The close is where finance turns that chaos into something the business can safely steer by.

Calling it a "process" can make it sound bureaucratic, like a ritual performed for its own sake. In a well-run company, the close is closer to a factory line: raw transactions go in, finished financials come out, and quality is controlled at multiple checkpoints. The output isn't just a P&L; it's a set of decisions the company will make with real consequences—pricing changes, hiring, spend cuts, bonus accruals, investor updates. Bad numbers don't merely embarrass you. They cause unforced errors.

The close has one central job: *freeze time*. You can't manage a business if the past keeps changing. The close draws a line under the month, applies the rules consistently, and produces statements that won't wobble every time a late invoice appears. That's why "final" matters. Not "kind of done." Not "close enough." Final means: reconciled, reviewed, and explainable.

There's a second, quieter job: *teach the organization what matters*. If your close routinely goes off the rails because of missing customer invoices, the company learns that billing discipline is optional. If your close is smooth because upstream teams hit deadlines and follow rules, the organization learns that accountability is real. Month-end behavior is culture, expressed in timestamps.

What actually happens during a close (and why it feels like a sprint)

A close is a controlled collision between two timelines. Timeline one is operational: customers buy, vendors deliver, payroll runs, ads spend, inventory moves. Timeline two is reporting: you need those events translated into accrual-based financial statements with clean classifications and support. The close is the moment those timelines are forced to align.

The work falls into four repeating movements, even if your company uses different labels.

First comes *capture*: make sure the period has the transactions it should have. That means all bank feeds imported, all credit card activity recorded, payroll posted, customer invoices issued, vendor bills entered (or accrued), and key subledgers updated. A surprising number of "finance problems" are simply missing data problems.

Second comes *adjust*: the accrual accounting layer that makes the month economically truthful. If you wait for every invoice to arrive before recognizing an expense, your reporting will be late and your trend lines will be nonsense. Adjustments are where you record what happened even when paperwork lags—accrued payroll, bonuses, commissions, prepaid amortization, revenue deferrals, inventory adjustments, depreciation, bad-debt reserves. Adjustments are not optional polish; they are the mechanism that converts transaction noise into period performance.

Third comes *reconcile*: prove that your balances tie to reality. This is the close's immune system. Reconciliations catch errors that otherwise masquerade as "business performance." A bank reconciliation proves cash. An accounts receivable reconciliation proves that the A/R aging ties to the general ledger and that the biggest past-due items have a story. A deferred revenue rollforward proves you didn't recognize revenue twice—or not at all. Inventory reconciliations prove you're not quietly bleeding through shrinkage or mis-costing.

Finally comes *review and lock*: variance analysis, management review, approval, and the formal act of saying, "This month is closed." Locking matters because it prevents retroactive edits that turn reporting into sand. If you have to reopen, you can—but reopening should feel like breaking the glass in an emergency, not like opening your fridge.

A close feels like a sprint because it is a sprint. But it should not feel like a fire drill. A sprint is planned; a fire drill is improvisation with smoke. The difference is preparation: tight cutoffs, clear ownership, pre-close work, and an expectation that upstream teams deliver inputs on time.

The cut-off: the moment you stop letting the month leak

Every close has a cut-off policy, whether you've admitted it or not. A cut-off is the line that decides what belongs in the month. If you don't draw the line intentionally, the line will be drawn by whoever yells loudest, which is an expensive way to run accounting.

Cut-offs exist because the world is asynchronous. Invoices arrive late. Shipping documents post after midnight. Expense reports come in whenever someone's guilt peaks. The point of a cut-off is not to pretend the world is tidy; it's to be consistently imperfect in a way that preserves trend integrity.

A good cut-off rule is simple enough that non-finance teams can follow it, and strict enough that the close doesn't become an endless waiting room. For example: all employee expenses must be submitted within two business days; after that, they hit next month unless they're above a defined materiality threshold. Or: revenue is recognized based on delivery evidence, not contract signature; if the evidence arrives after the cut-off, the revenue waits.

Materiality is the grown-up word that gives you permission to be pragmatic. Not every late $200 bill deserves to delay reporting and burn a day of leadership attention. The close is where finance learns to say, "We will be precise where it changes decisions, and fast where it doesn't." That's not sloppiness. That's prioritization.

Adjusting entries: the small journal entries that prevent big lies

Adjusting entries sound like accounting clerical work until you realize they are the difference between truth and timing. A company without a disciplined adjusting-entry routine will report results that depend on vendor billing patterns, payroll calendars, and how aggressively people submit expenses. That's not insight; that's administrative randomness.

Accrued payroll is the classic example because everyone understands it. Employees earn wages every day, not just on payday. If you only record payroll when cash leaves, some months will look artificially profitable and others artificially unprofitable, based purely on calendar quirks. Accrue the earned wages in the period they were earned. Reverse the accrual next period when the payroll hits. Suddenly your labor cost trend means something.

Revenue deferrals are the mirror image: cash collected isn't always revenue earned. If you sell annual contracts and recognize them all in month one, your P&L becomes a carnival ride—fun until you try to drive with it. Deferral schedules aren't "accounting for accounting's sake." They are how you report the economics of the promises you've made.

Then there are estimates—bad debt, returns, warranty, bonuses. Estimates make some operators uncomfortable because they feel subjective. But refusing to estimate doesn't remove subjectivity; it just pushes the uncertainty into future periods, where it arrives as "surprise" expense. A disciplined close forces the company to look uncertainty in the eye every month and update its best view of reality.

The best adjusting entries share a trait: they are repeatable. If your team is inventing brand-new accrual logic each month, you don't have a close process; you have improvisational theater. Repeatable entries can be templated, scheduled, and reviewed quickly. That frees human brainpower for the real work: interpreting the results.

Reconciliations: your numbers are guilty until proven innocent

Reconciliations are not chores. They are proofs. When the ledger says you have $2.3M in cash, "trust me" is not an acceptable supporting schedule. When the ledger says accounts receivable is $4.8M, you should be able to point to an aging report and explain the top ten balances without blinking.

A reconciliation has three elements: a source of truth, a tie-out, and an explanation for differences. The bank statement is a source of truth. The general ledger is a claim. The reconciliation is the bridge. Differences might be timing (outstanding checks), errors (double-posted receipts), or fraud (rare, but the whole point is to notice if it happens). You're not only validating the number; you're validating that your process would *catch* a problem if one appeared.

Some reconciliations are especially high-leverage because they touch many downstream lines. Cash, of course. Accounts receivable and deferred revenue, because they are the plumbing between sales activity and reported performance. Payroll liabilities, because missing or misclassified payroll entries create both financial and compliance risk. Fixed assets, because capitalization errors are slow-moving but persistent—like a small leak that ruins a ceiling over time.

A reconciliation culture also protects the business from a very specific failure mode: confusing "balance sheet junk" with "immaterial." Balance sheet accounts often get neglected because they don't hit EBITDA directly. That's how you end up with old accruals that never reverse, prepaid accounts that never amortize, clearing accounts that become graveyards, and deferred revenue schedules that quietly drift. Those issues don't stay on the balance sheet forever; they eventually spill into the P&L in the worst possible month.

Closing entries: resetting the scoreboard without erasing history

Once the month is adjusted and reconciled, accounting does something that feels ceremonial but is deeply practical: it closes temporary accounts. Revenue and expense accounts are "temporary" because they measure a period. At period end, their balances are transferred into retained earnings (through whatever formal closing mechanism your system uses), and then they reset to zero for the new period. Balance sheet accounts are "permanent" because they carry forward; they represent what the business has and owes at the boundary between months.

If you've never watched this happen, it can feel abstract. A simple mental model helps: imagine a whiteboard where you tally the month's wins and losses. At month-end, you circle the net result and write it into a running total on the side. Then you wipe the monthly tallies clean to start fresh, keeping the running total. That running total is retained earnings. The wipe is the closing entry.

This reset is what makes monthly reporting comparable. Without it, revenue and expense accounts would accumulate forever, and you'd lose the ability to answer basic questions like, "How did we do *this month*?" Closing entries are the technical step that makes the timebox real.

A practical control tied to closing entries is the post-closing trial balance. It's an unglamorous report with a glamorous purpose: after closing, only balance sheet accounts should have balances. If revenue or expense accounts still show amounts, something didn't close correctly. And if debits don't equal credits, you don't have financial statements-you have a math error wearing a suit.

Speed vs. accuracy: why Day 5 beats Day 25 (almost every time)

The most common close mistake isn't an incorrect journal entry. It's treating perfection as the goal. In operating companies, information has a shelf life. A flawless set of numbers delivered when the decisions have already been made is like a fire extinguisher that arrives after the building is ash.

A "fast close" doesn't mean reckless. It means designing the process so that the biggest drivers are accurate quickly, and the tiny details don't hold the business hostage. The best teams build a close that is *predictably* fast: day-by-day cadence, hard deadlines, standard accruals, and clear roles. Predictability reduces stress because everyone knows what happens when, and bottlenecks become visible enough to fix.

Accuracy still matters, but it needs a definition that reflects decision-making. The goal is "accurate enough that leaders won't make the wrong call." That often implies focusing effort on the lines that actually move decisions: revenue, gross margin, payroll, major vendor spend, key working-capital balances, and cash. A $1,200 misclassification inside office supplies is not worth breaking your team for. A $120,000 misstatement in revenue timing absolutely is.

This is also where good finance teams use a two-stage mindset: preliminary and final. Preliminary numbers give leaders directional insight early (often with known, bounded estimates). Final numbers incorporate late-arriving information and complete reconciliations. The key is that the differences between preliminary and final should be explainable and shrinking over time. If "prelim" and "final" feel like two different realities, your upstream processes are failing.

Close as an operating system: calendars, ownership, and the tyranny of small delays

The close is rarely delayed by one big problem. It's delayed by twenty small ones: missing approvals, unclear account ownership, a vendor portal change, a new sales comp plan that finance hears about after commissions are paid, a warehouse count that slips a day. Those delays compound, and finance becomes the bottleneck for issues it didn't create.

A strong close process assigns owners to balance sheet accounts the way an engineering team assigns owners to production services. "Owned" means someone is responsible for understanding what moves the account, reconciling it, and explaining it. Ownership turns the close from a vague collective responsibility into a series of accountable deliverables.

Calendars matter for the same reason airline schedules matter: you can't run a complex system with everyone improvising. A reporting calendar defines deadlines for invoice submission, purchase approvals, billing, inventory counts, and review meetings. It makes the close a company-wide routine, not finance's private suffering. When non-finance leaders know the cut-off dates, they start planning around them, and the system stops relying on heroic reminders.

Automation helps, but only after the logic is clear. A messy process automated is just a faster mess. The highest ROI automation is usually mundane: bank feed rules, recurring journal entries, standardized accrual templates, approval workflows, and system integrations between billing, payroll, and the general ledger. Each automation removes a manual touchpoint, and every manual touchpoint is a place where time and errors breed.

Why the close is inseparable from control (even in private companies)

Public companies talk about internal control over financial reporting because they have to. Private companies should care because reality doesn't give discounts for being private. A sloppy close is not merely an accounting inconvenience; it's an enterprise risk. If your numbers can't be trusted, you can't price properly, you can't forecast cash properly, you can't measure performance properly, and you can't raise capital on good terms when you need it.

Close controls are the foundation: reconciliations performed and reviewed, approvals documented, access limited, changes tracked, and key estimates supported. Controls can sound like handcuffs, but good ones feel like guardrails: they let the team move faster without driving off a cliff. The goal is reasonable assurance, not paranoia. You're designing a system that makes it hard to make big mistakes and easy to catch small ones early.

One useful mental model is distinguishing between a control that is well-designed and one that is operating well. A bank reconciliation control can be perfectly designed on paper ("reconcile monthly, reviewer signs off") and still fail in operation if it's done late, done sloppily, or reviewed by someone who doesn't understand what they're reviewing. The close exposes operational control weaknesses quickly because the deadlines are unforgiving.

A close that scales: what changes as the company grows

Early-stage companies often close with a mix of spreadsheets, late nights, and "we'll fix it later." That can work for a while, but it's a debt that compounds. As transaction volume grows and stakeholders demand tighter reporting, the cost of improvisation becomes brutal. The close starts eating your best people. Finance becomes reactive. Errors slip through because everyone is exhausted.

Scaling the close isn't about adding bodies first; it's about adding structure. Standardize recurring entries. Define a materiality threshold. Lock cut-offs. Implement account ownership. Create a close checklist that is actually used. Then automate what repeats. Headcount should be the last lever, not the first—otherwise you're hiring people to carry water in buckets instead of fixing the pipes.

A mature close also becomes more analytical. Once the mechanics are predictable, finance can spend more time on variance analysis, trend explanations, and operational insights. That's the moment the close stops being "getting numbers out" and starts being "making numbers useful."

Actionable takeaway: Build a one-page close checklist and run it for the next two cycles. For each major balance sheet account (cash, A/R, A/P, deferred revenue, payroll liabilities), assign an owner and require three artifacts by a fixed day: (1) a reconciliation that ties to an external or subledger source, (2) a short explanation of material movements, and (3) evidence of review (a signature, comment, or approval). Then measure two things: days to close and number of post-close adjustments. If days to close don't improve and adjustments don't fall, you don't have a close problem—you have an upstream discipline problem, and the checklist will tell you exactly where it lives.

2.4 Trust and Verification

A spreadsheet can be perfectly formatted and still be fiction. The lie doesn't have to be dramatic; it can be boring, incremental, and well-intentioned. A reimbursable expense coded to the wrong department. A duplicate vendor set up with a slightly different name. A discount buried in "other." A revenue schedule that drifts because nobody owns the deferred revenue rollforward. Over time, those small mis-truths don't stay small. They accumulate into confident decisions made on shaky ground.

Trust in financial reporting isn't a vibe. It's an engineered outcome. When leaders say, "I trust the numbers," what they're really saying is, "I trust the system that produced them." That system is a mix of process, people, and safeguards—what accountants call internal controls. Controls are not there to slow the business down. They exist so the business can move fast *without* driving blindfolded.

Verification is the other half of the equation. A control says, "This is how we prevent or catch errors." Verification says, "Show me it worked this month." A close checklist without evidence is a wish. A reconciliation without review is a diary entry. In well-run finance organizations, the evidence is boring, consistent, and easy to pull up—because the real goal is not to win an audit; it's to prevent a bad number from becoming a bad decision.

Internal controls: the guardrails that keep the scoreboard honest

Internal control over financial reporting (ICFR) is a mouthful, but the idea is simple: build a system that provides reasonable assurance your financial statements are reliable and prepared according to the rules you've chosen. "Reasonable" matters. No one is promising perfection. The promise is that material mistakes-errors large enough to change decisions-are unlikely, and if they start to happen, they'll be detected in time to fix them.

Controls come in two flavors that behave differently under pressure: preventive and detective. Preventive controls stop bad things from happening in the first place (approval workflows, system access restrictions, three-way match for purchases). Detective controls find bad things after they happen but before they metastasize (bank reconciliations, variance analysis, management review, exception reports). Most companies need both. A preventive-only mindset turns into bureaucracy. A detective-only mindset turns into constant cleanup.

A useful way to make controls feel less abstract is to anchor them to the moments where money changes hands or obligations are created:

- When you *pay* someone: controls prevent unauthorized payments and duplicate payments.
- When you *bill* someone: controls prevent missed invoices and incorrect revenue recognition.
- When you *change* master data (vendors, customers, bank details): controls prevent fraud's favorite trick—rerouting cash.
- When you *estimate* and *accrue*: controls prevent optimism from masquerading as accounting.

Those are the pressure points. That's where you invest rigor.

Control design also forces a business to confront an uncomfortable truth: most fraud is not a movie plot. It's opportunity plus familiarity plus a long time without anyone checking. The goal isn't to assume everyone is dishonest. The goal is to assume everyone is human-busy, sometimes careless, occasionally tempted-and to build a system where a single person can't quietly do too much without leaving footprints.

Segregation of duties: the simplest anti-fraud idea that's still widely violated

Segregation of duties is the principle that no single person should control all parts of a transaction. In its cleanest form, you separate:

- *Authorization*: approving the transaction (who can commit the company).
- *Custody*: holding or moving the asset (who can move cash or inventory).
- *Record keeping*: recording the transaction (who can create the accounting history).

When one person holds two of these, the system becomes vulnerable. Not because that person is necessarily untrustworthy, but because the system now relies on a single point of failure.

A classic small-company violation is also the most understandable: the same person sets up vendors, enters bills, and processes payments. It happens because the company is lean and everyone is trying to move fast. But it's the accounting version of leaving your house keys under the doormat and then being surprised when someone finds them.

Segregation of duties isn't always achieved with headcount. You can split responsibilities using workflow approvals, dual signatures, locked-down bank portals, and role-based permissions in your accounting system. You can require independent review of vendor changes. You can restrict who can edit bank account fields. You can enforce that the person who reconciles the bank account cannot initiate payments. In practice, good segregation is often a patchwork of system settings and process discipline rather than three different people in three different rooms.

A strong finance leader treats segregation of duties like an architecture problem: where are the dangerous combinations, and what's the least painful way to break them? The answer differs by company size, but the principle doesn't.

Controls that matter most in the real world (not in theory)

Some controls sound important but catch almost nothing. Others feel mundane and catch everything. The high-impact controls tend to share two traits: they tie to an independent source of truth, and they happen on a schedule the business can actually maintain.

Cash controls are the obvious starting point. Bank reconciliations aren't glamorous, but they are powerful because they force the ledger to agree with an external record. If the bank rec is late, sloppy, or not reviewed, your finance organization is essentially choosing to be surprised later.

Revenue controls deserve the same seriousness because revenue is where incentives are strongest. That doesn't mean distrusting Sales; it means acknowledging that humans respond to targets. Strong revenue controls include documented revenue recognition policies, system-driven revenue schedules when possible, and a monthly tie-out of billed amounts, deferred revenue movements, and recognized revenue. If you can't explain the month's revenue in plain language, you don't have revenue controls-you have a revenue hope strategy.

Vendor and payment controls are where many frauds begin and where many expensive errors hide. The simplest upgrade with outsized impact is controlling changes to vendor master data and bank details. Fraudsters rarely "hack the general ledger." They impersonate a vendor, submit new banking instructions, and hope your process is weak enough to comply. A call-back procedure (to a known number, not the email signature) and approval for changes is dull, repetitive, and deeply effective.

Journal entry controls are another underappreciated area. Journal entries are powerful because they can move financial results without a corresponding operational transaction. That's necessary for accrual accounting, but it's also an attack surface. A healthy control environment limits who can post manual entries, requires support for entries above a threshold, and includes review of unusual entries-especially late in the period.

Inventory and cost controls matter whenever physical goods or usage-based costs are meaningful. Shrinkage, miscounts, incorrect standard costs, and cut-off errors can quietly distort margins for months. The fix is rarely "count more." It's designing a repeatable method for counts, reconciling perpetual records to physical reality, and investigating variances like you mean it.

All of these controls become dramatically more valuable when they are integrated into the close. A control that happens "when someone remembers" is not a control. A control embedded in a five-day close cadence is a habit the company can keep.

Materiality: why small errors can be big problems

A common misunderstanding is that materiality is a math rule, like "anything under 5% doesn't matter." In reality, materiality is contextual. A small misstatement can be material because of *where* it lands (turning a loss into a profit), *what* it implies (masking a trend), *why* it happened (intentional manipulation), or *who* cares (a covenant, a regulatory metric, a key KPI). Materiality is about whether the misstatement could influence decisions, not whether it clears an arbitrary percentage hurdle.

This matters because it shapes how you design controls and how you respond to errors. If a manager says, "It's only $20k," the right response isn't automatically "ignore it." The right response is, "Relative to what, and what story does it change?" A $20k error in office supplies is noise. A $20k error in revenue at quarter-end that flips a growth metric is a governance issue.

Materiality also has a psychological trap: if you tolerate small intentional misstatements, you teach the organization that the numbers are negotiable. That's when finance stops being a scorekeeper and becomes a PR function. Once that happens, talented operators stop trusting reporting and start running their own shadow spreadsheets, which is how companies lose a single version of the truth.

Audits and assurance: what ``reasonable assurance" actually feels like

Audits are often described as verification, but they're really structured skepticism. The auditor's opinion is not "these numbers are perfect." It's "these statements are fairly presented, in all material respects, under the applicable framework." That wording is careful on purpose. The world is messy, estimates exist, and humans make mistakes. The audit is designed to reduce the risk that material errors slip through undetected.

The most valuable way to think about an audit is not as a punishment but as a forcing function. If your reconciliations are weak, your revenue recognition is fuzzy, or your close relies on heroics, an audit makes that expensive. The pain is not the audit itself; the pain is the gap between how disciplined the business needs to be and how disciplined it currently is.

There's also an important distinction that surprises many leaders: a company's internal controls can be deemed ineffective because of a material weakness even if the financial statements are not yet materially misstated. Controls are about the *ability* to prevent or detect misstatements on a timely basis. In other words, you can get the right answer by luck or by late fixes and still have a control problem. A CFO should care because luck is not a scalable reporting strategy.

Management representation letters—those formal statements management signs for auditors—reinforce a philosophical point: management owns the financial statements. Auditors provide assurance, not ownership. A strong finance organization behaves that way even without an audit requirement: finance owns the truth, and verification is how it keeps that ownership honest.

KPIs: the fastest way to destroy trust (or earn it)

Nothing erodes confidence faster than two executives showing up to the same meeting with two versions of "ARR" or "gross margin." People often assume the disagreement is about performance. More often it's about definitions. One report includes discounts; another excludes them. One treats implementation fees as revenue; another treats them as pass-through. One counts churn by logo; another counts it by dollars. Everyone is technically "right," and the organization becomes collectively confused.

A KPI is only as useful as its definition and its data lineage. "Lineage" means you can trace the number from the dashboard back to the system of record, through transformations, to the final calculation. When lineage is unclear, metrics become political. People choose the version that flatters their function. The KPI stops being a tool and becomes a weapon.

A robust KPI definition includes:

- The formula (what's included and excluded).
- The timing rule (when it's counted).
- The source systems (billing, CRM, general ledger, data warehouse).
- The owner (who is responsible for accuracy and changes).
- The change control (how the definition evolves without breaking comparability).

That may sound heavy, but it's lighter than the alternative: endless debates, broken trust, and decisions delayed because nobody believes the scoreboard.

The most important KPI design decision is whether the metric is meant to be *accounting-grade* or *management-grade*. Accounting-grade metrics tie tightly to the general ledger and follow formal recognition rules. Management-grade metrics can be more forward-looking or operational (pipeline coverage, retention cohorts, activation rates). Both are valuable, but they must be labeled honestly. Trouble begins when a management metric is presented with the confidence of audited financials, or when an accounting metric is expected to behave like a real-time operational dashboard.

One practical move that elevates reporting quality overnight is building KPI reconciliation habits. If you report bookings, reconcile bookings to billings and deferred revenue movements. If you report ARR, reconcile ARR movement to invoicing and churn events, and document differences (timing, contract structure, non-recurring items).

Reconciliations aren't just for balance sheet accounts; they're for truth.

Designing a control environment people will actually follow

Controls fail for predictable reasons: they're unclear, they're too slow, they're too easy to bypass, or they rely on one overworked person's memory. A control that makes everyone's life harder will be bypassed with creative enthusiasm. A control that is embedded into tools and workflows becomes invisible—and therefore sustainable.

The secret is to align controls with incentives and reality. If sales commissions depend on deals being set up correctly in the system, the system gets used correctly. If purchase approvals are quick and consistent, people stop treating them as optional. If expense policies are reasonable, employees stop playing games. Controls are as much behavioral design as they are accounting design.

It also helps to separate "control" from "control theater." Control theater is when you add checklists, signatures, and meetings that produce the feeling of rigor without actually reducing risk. Real controls either prevent a plausible error/fraud scenario or detect it quickly. If you can't name the scenario, you probably have theater.

A mature organization periodically performs a risk scan: where could a material misstatement come from? New revenue streams, new systems, rapid hiring, complex estimates, and significant incentives are classic risk multipliers. Controls should evolve as the business evolves. The control environment that worked at $5M of revenue will break at $50M—not because people got worse, but because the company got faster and more complex.

Actionable takeaway: Pick one high-risk flow—*vendor payments* or *revenue recognition*—and run a "control walk" this week. Write down every step from initiation to recording, and for each step answer two questions: "Who can do this?" and "Who checks this?" If you find any step where the same person can initiate, approve, and record without independent review, break the chain immediately using system permissions or workflow approvals. Then choose one KPI your leadership argues about, publish a one-page definition with an owner and a data source, and freeze that definition for a quarter. You'll feel the organization exhale the next time the numbers hit the table.

Chapter 3

The Lifeblood: Cash, Treasury, and Working Capital

Profit is an opinion; cash is a fact. You can survive for years without a profit, but you will die in days without cash. This chapter moves beyond accounting theory to the gritty reality of bank balances, payment runs, and survival. We will build your 'cash flashlight' to see around corners, tighten the plumbing of your treasury operations to prevent leaks and fraud, and learn how to unlock the hidden millions trapped inside your own supply chain.

3.1 Liquidity and Survival: The Oxygen of Business

A company can be "successful" and still die on a Tuesday.

Monday looks fine: customers love the product, the pipeline is healthy, the board deck is optimistic. Tuesday is payroll. The bank balance is not optimistic. Somewhere between those two facts lives the brutal truth of corporate mortality: businesses don't go bankrupt when they run out of ideas; they go bankrupt when they run out of cash at the wrong moment.

Liquidity is that moment-management. It's the difference between having value *somewhere* in the enterprise and having spendable dollars *where* they're needed, *when* they're needed. Solvency is a balance sheet concept. Liquidity is a calendar problem with teeth.

Solvent on paper, suffocating in practice

Picture a business with $20 million of equity value in a fundraising memo. It owns equipment. It has a recognizable brand. Its customers renew. The founders speak in confident sentences.

Now picture the same business with $180,000 in the operating account, $410,000 of payroll and taxes due in nine days, and a customer that pays net-60 but *means* net-75. That company is not thinking about "valuation." It's thinking about whether it can keep the lights on without violating wage laws, vendor terms, or basic decency.

Liquidity is the ability to meet obligations as they come due *without* doing something desperate: missing payroll, stiffing a critical supplier, defaulting on a covenant, issuing equity at a punitive price, or taking a loan whose true interest rate is shame.

Two traps cause smart teams to underestimate this:

- *The accrual trap:* the income statement says you made money, so you assume cash increased. Accrual accounting recognizes revenue when earned and expenses when incurred. Your bank account recognizes only what cleared.
- *The asset trap:* you assume your assets are there to save you. Many assets are real and valuable—they just aren't liquid on a Thursday afternoon when the payroll file must be released.

A CFO earns credibility by being calmly allergic to these traps. Not paranoid. Not pessimistic. Just structurally incapable of confusing "good business" with "cash today."

Liquidity has three dimensions: amount, timing, and certainty

Most liquidity conversations start with the wrong question: "How much cash do we have?"

The better question is: "How much cash will we have, on each critical date, with what level of confidence?" A dollar that arrives tomorrow is not a dollar today. A dollar that arrives "probably" is not a dollar you can safely spend.

Liquidity has three dimensions:

- *Amount:* the size of the buffer.
- *Timing:* whether the buffer exists on the dates that matter (payroll, tax remittances, rent, interest, vendor runs).
- *Certainty:* whether expected cash is contractual, habitual, or hopeful.

When liquidity fails, it's rarely because leadership didn't know the *amount* of cash on hand. It fails because leadership misjudged timing and certainty: the customer paid late, the chargeback hit, the bank took an extra day to release funds, the quarter-end "one-time" payment didn't repeat, the promised bridge financing didn't bridge.

That's why mature finance teams talk about liquidity like engineers talk about load: not as an average, but as a stress case.

The existential math: burn and runway

Liquidity management becomes practical when it becomes arithmetic.

Cash burn is the net cash outflow over a period—how much the bank balance declines, after receipts and disbursements, over a month (or week). *Runway* is how long you can keep operating before you hit a minimum cash threshold, given your burn.

The simplest (and often most useful) definitions:

$$\text{Net Burn (per month)} = \text{Cash Outflows} - \text{Cash Inflows}$$

$$\text{Runway (months)} = \frac{\text{Starting Cash} - \text{Minimum Cash Floor}}{\text{Net Burn per month}}$$

The *minimum cash floor* is where real CFOs start arguing, because it forces you to define what "zero" actually means. If you run payroll every two weeks, "zero" is not literally $0. It might be one payroll plus payroll taxes plus a small margin for error. If you are in retail with daily card settlement, your floor might be lower. If you are in construction with lumpy draws and big vendor checks, your floor might be higher.

A useful discipline: define *Cash Zero* as the first day you cannot meet obligations (not the day the account hits zero). That day arrives earlier than your spreadsheet wants to admit.

A runway calculation that respects reality

Consider a service business with:

- Starting unrestricted cash: $3.2M
- Expected monthly receipts: $1.1M
- Expected monthly disbursements: $1.6M (payroll, rent, vendors, taxes)
- Minimum cash floor: $0.6M (one payroll cycle + tax cushion + operating slack)

Net burn = $1.6M - $1.1M = $0.5M per month.

Runway = ($3.2M - $0.6M) / $0.5M = 5.2 months.

That seems clear. It is also incomplete, because it hides two nasty truths:

- Receipts are not a metronome. If 40% of receipts come from two customers, your "monthly receipts" is a story, not a fact.
- Disbursements are sticky. Costs often don't fall smoothly. They fall in cliffs: you either keep a team or you don't; you either renew the software suite or rip it out.

A runway number is not a prediction. It's a *conversation starter* that forces you to specify assumptions and then interrogate them like they owe you money.

Base case vs. stress case: the CFO's two-lens view

Every leadership team has a base case. The base case is how the world works when you are right about yourself and the world is normal.

A CFO also carries a stress case—not because doom is fashionable, but because the world is allergic to your model. Stress cases are where you find out if you're managing a business or merely narrating one.

A practical stress case begins with a simple question:

> What happens if revenue hits zero tomorrow?

For most businesses, revenue won't hit zero. The point is not realism; the point is to expose the fixed cost structure and the time it takes to reduce cash outflow when forced. Stress testing is less about the specific shock and more about your *response latency*. If sales fall 30%, how fast can the organization actually cut spending 20%? A week? A month? A quarter? In many companies, the honest answer is: "Not before the cash is gone."

Runway under stress: a more honest numerator and denominator

Stress runway is different in two ways:

- *The numerator shrinks:* some cash is not truly available. It may be restricted, pledged, or needed as collateral. Even unrestricted cash may not be usable if it triggers covenant issues or violates practical obligations.
- *The denominator grows:* receipts become less certain and often slow down. Disbursements often rise before they fall (severance, contract termination fees, catch-up tax payments).

A stress runway calculation asks you to reframe both:

$$\text{Available Cash} = \text{Unrestricted Cash} - \text{Do-not-touch Reserves}$$

Stress Burn = Committed Outflows + Likely Shock Costs – Highly Certain Inflows

The phrase *highly certain* is doing a lot of work. "Customer said they'd pay" is not highly certain. "ACH is scheduled and has historically cleared" is closer. "Contractual monthly minimums from 200 customers with card-on-file" is better. The goal is not cynicism; it's survival-weighted truth.

A stress test that turns abstract fear into specific dates

Take the same company with $3.2M of cash and a $0.6M floor.

Stress assumptions:

- Receipts fall 35% immediately (from $1.1M to $0.715M) due to churn and slower collections.
- Disbursements fall only 10% in the first month (from $1.6M to $1.44M) because payroll and contracts are sticky.
- One-time shock costs: $0.35M (legal, retention, severance, expedited collections work, termination fees).

Stress burn in month 1 becomes:

Monthly net burn = $1.44M - $0.715M = $0.725M, plus $0.35M shock = $1.075M.

Now runway is:

($3.2M - $0.6M) / $1.075M = 2.4 months.

That's not a rounding error. That's the difference between having time to make good decisions and having time only to make fast ones. It's also the difference between negotiating from strength and negotiating while your counterparty smells the smoke.

The line chart that forces adult conversations

The most useful liquidity visual isn't a ratio. It's a trajectory: cash over time under two scenarios, with the "Cash Zero" date marked like a cliff edge.

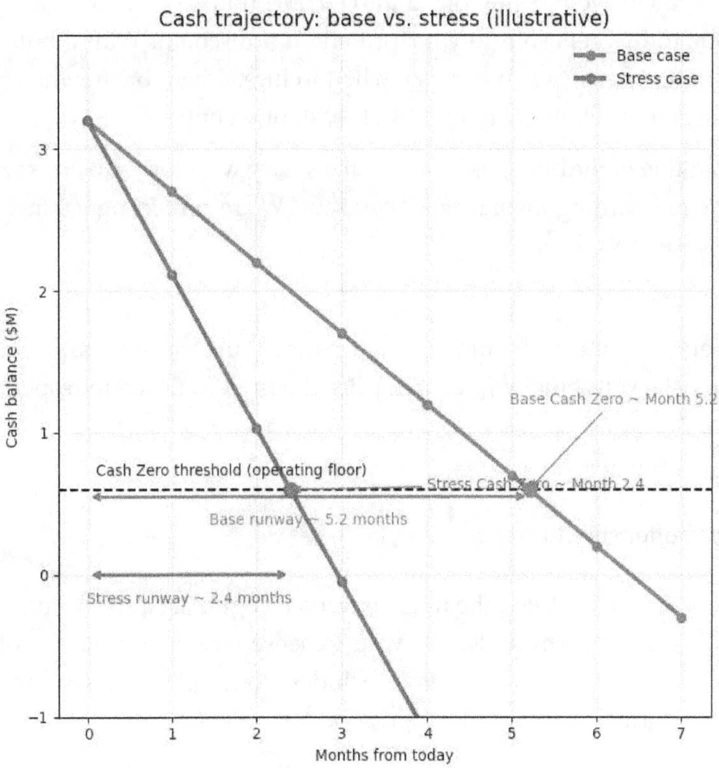

A line like that changes behavior. It turns vague optimism into a date on the calendar. Teams can argue with a philosophy; they can't argue with a payroll file.

The rainy day buffer: not hoarding, insurance

Plenty of leaders hate idle cash because it feels like laziness. Cash earns little. It creates temptation. It makes you look conservative.

That's a dangerous aesthetic preference masquerading as strategy.

A liquidity buffer is not a sign you lack ambition. It's the premium you pay to avoid being forced into terrible trades. Without a buffer, your future gets sold to your present at a discount. With a buffer, you can choose when to raise, when to hire, when to cut, when to renegotiate. Optionality is not free; cash often buys it.

Thinking of the buffer as *insurance* fixes the psychology. No one says, "We're hoarding insurance." They say, "We're protecting against a loss we can't absorb."

So what is "adequate"?

There is no universal number, but there is a universal method: define the risks you actually face, then price the time you need to respond.

Sizing the buffer by response time, not vibes

The buffer should cover:

- *Response latency:* the time between recognizing a shock and reducing cash outflow. If your expense base is mostly payroll, your true response time includes notice periods, severance norms, and the reality that good people need time and dignity.
- *Cashflow variability:* how lumpy your receipts are, and how late they can be without triggering a crisis.
- *Concentration risk:* if a small number of customers represent a large share of receipts, your buffer must be bigger than your average month suggests.
- *One-time shock costs:* the costs that appear when you *do* respond (severance, legal, contract termination, expedited collections, audit and tax complexity).

A sober way to size a buffer is to pick a stress scenario you can defend in the boardroom and then ensure you have enough cash to execute a plan under that scenario without improvising.

For example: "If receipts fall 25% for two quarters and collections slow by 15 days, we can maintain operations and execute a cost reset in 60 days without missing payroll or violating covenants." That sentence contains more operational truth than any generic "six months of cash" rule—and it forces you to quantify the 60 days.

Minimum cash floors: the CFO's non-negotiables

A minimum cash floor is not just a number; it's a policy. It answers: "At what point do we stop pretending we have choices?"

A strong floor policy has three layers:

1. *Hard obligations:* payroll, payroll taxes, sales/VAT remittances, debt service, rent where nonpayment triggers eviction or default.
2. *Operational continuity:* the expenses required to keep producing and delivering the product in the near term (critical vendors, cloud infrastructure, insurance).
3. *Error margin:* because banking systems have cutoffs, wires get held, card settlements can be delayed, and humans make mistakes. The error margin keeps a minor delay from becoming a major incident.

What belongs *outside* the floor? Nice-to-haves. Experiments. Discretionary marketing. Travel. Hiring that can be postponed. Anything that feels painful but not fatal.

A floor is only useful if it triggers behavior. If crossing the floor doesn't cause immediate restrictions, it wasn't a floor. It was a suggestion.

Liquidity is also about access: the difference between cash and cash *plus*

Some companies obsess over cash balances and ignore access to liquidity: credit facilities, invoice financing, customer prepayments, equity lines, or simply the ability to reduce working capital quickly. Access is valuable, but it has fine print.

A revolver is not cash. It is a bank's promise—contingent on covenants, borrowing base calculations, and the bank's comfort with your trend line. The moment you need liquidity most is often the moment access tightens.

A practical CFO treats external liquidity like a fire extinguisher: it should be serviced, tested, and reachable. And if the building is already on fire, it's too late to read the instructions.

Stress testing isn't a spreadsheet exercise; it's a decision rehearsal

The best stress tests feel a little like disaster drills. Not because you expect the disaster, but because you don't want the first time you think about it to be the day it happens.

A liquidity stress test should force explicit answers to uncomfortable questions:

- *What costs are truly fixed for the next 30, 60, 90 days?* Not "fixed in theory"—fixed in contract, in culture, or in operational necessity.
- *What levers exist, and how fast do they work?* Cutting software renewals might save in 45 days; cutting paid acquisition might save next week; renegotiating rent might save next quarter.
- *Who decides, and how quickly?* The best cost plan in the world is useless if approvals require three committees and a board meeting.

- *What are the second-order effects?* A hiring freeze might protect cash but increase churn if support quality drops. A vendor stretch might conserve cash but disrupt supply and revenue. Liquidity plans that ignore second-order effects are just delayed bankruptcy in nicer formatting.

A good stress test produces not only a runway number, but a *playbook*: triggers, actions, owners, and timeframes. It is hard to overstate how comforting it is in a crisis to already know what you'll do—and to have pre-aligned on what "crossing the line" means.

The quiet killer: liquidity mirages

Some cash looks real until you touch it.

Common mirages include:

- *Receivables you can't collect quickly:* disputed invoices, customers with chronic lateness, large balances tied to completion criteria, or customers who pay only after their own funding events.
- *Inventory as "cash on shelves":* inventory is cash you already spent, now wearing a costume. Turning it back into cash requires demand, time, and often discounting.
- *One-time inflows mistaken for a trend:* tax refunds, a settlement, a delayed customer payment that makes one month look heroic.
- *Restricted cash:* funds that exist but are encumbered by contractual terms, escrow agreements, or regulatory restrictions.

The antidote is a liquidity lens that asks, relentlessly: "When can we convert this to cash, at what discount, with what confidence?"

If the answer includes the words "should" or "eventually," it's not liquidity. It's hope with a due date.

How leaders accidentally sabotage liquidity (and how CFOs prevent it)

Liquidity crises are often self-inflicted, not because leaders are reckless, but because they're doing what leaders are trained to do: invest, grow, push.

The sabotage patterns are predictable:

- *Hiring as a default response to problems.* Payroll is the stickiest expense. It is also the most human. Every hire is a future conversation you may not want to have under stress.
- *Overcommitting to annual contracts.* Annual prepayments and minimums feel like "locking in a deal" until the deal locks you in.
- *Treating liquidity as finance's problem.* Liquidity is an operating system. Sales, ops, product, and procurement all touch it. If finance is the only team who cares, finance will eventually be the only team left.

A CFO doesn't fix this with lectures. They fix it with operating rhythm: a runway metric everyone sees, a clear cash floor policy, and a discipline of testing assumptions when they are cheap to test.

Actionable takeaway: a 48-hour liquidity drill that changes your next 90 days

Do this once, quickly, and you will feel the company's true liquidity profile in your bones.

1. *Define Cash Zero.* Pick a minimum cash floor that reflects obligations (next payroll + taxes + critical vendors + error margin). Write the number down.
2. *Calculate base burn and stress burn.* Base: last 3 months average net cash outflow. Stress: assume receipts down 25–35%, collec-

tions slower, and include one-time shock costs (severance/termination/legal).

3. *Compute two runways to Cash Zero.* One for base, one for stress. Convert them into dates on the calendar.
4. *Write three triggers with owners.* Example: "If stress runway < 12 weeks, freeze nonessential hiring (People Ops owner). If < 10 weeks, cut discretionary spend by $X (department owners). If < 8 weeks, initiate cost reset plan and vendor renegotiations (CFO/COO owners)."

If you can't complete this drill in 48 hours, the problem isn't your spreadsheet. It's visibility. And if you *can* complete it in 48 hours, you've just purchased something rare: the ability to make decisions before your bank account makes them for you.

3.2 The 13-Week Crystal Ball: Predicting Cash Flow

A cash forecast is the closest thing business has to a weather report that actually matters. Not the cinematic kind with dramatic music and swirling storm graphics—the useful kind that tells you, plainly, whether you should cancel the picnic.

The problem with most financial reporting is that it's polite. The income statement is tidy; the balance sheet is symmetrical; the cash flow statement is logical in retrospect. Your bank account is not polite. It is a live wire. It doesn't care that revenue was "earned," only that money *arrived*. It doesn't care that an expense is "non-cash" if the vendor still expects to be paid.

A 13-week cash forecast is how a CFO stops being surprised. Thirteen weeks is long enough to see trouble coming and short enough to be specific. It fits inside a quarter (the unit of executive attention) while refusing to hide behind quarterly averages. It also forces the right kind of discipline: if you can't predict cash for the next 91 days with some accuracy, the business isn't being managed—it's being narrated.

Why 13 weeks works when annual budgets fail

Annual budgets are aspirational; they assume the organization will behave. Cash doesn't reward aspiration. It rewards *timing*.

Thirteen weeks is the horizon where timing becomes visible. Payroll cycles repeat. Rent hits like clockwork. Tax remittances appear on the calendar with the cold inevitability of seasons. Customer payments, too, have patterns: some clients pay on the 1st, some on the 15th, some only after three reminder emails and a phone call that feels like swallowing gravel.

There's also a psychological reason 13 weeks works: it produces a forecast you can argue with. Not in the abstract, but line by line. "Is that $240k from Customer A really coming in Week 4? Who confirmed? What happens if it slides to Week 6?" Those are adult questions. They are also the questions that keep companies alive.

Direct vs. indirect: the difference between counting cash and explaining it

Finance teams love the indirect method because it reconciles neatly to accrual accounting: start with net income, add back non-cash items, adjust for working capital changes, and arrive at cash from operations. It's elegant. It's also the wrong tool when you're trying to avoid a short-term cash crunch.

The direct method is deliberately unglamorous. It asks: *What cash will hit the bank, and what cash will leave the bank, each week?* That's it. No accounting poetry. Just receipts and disbursements.

If the indirect method is a post-game interview—"Here's why we won"—the direct method is a play clock. It tells you how many seconds you have before the next snap, and whether you can afford to call something fancy.

When a company is inside a tight liquidity window, the indirect method can be actively misleading. A business can show improving EBITDA while cash still falls, simply because collections lag, inventory builds, or payables get cleaned up. The direct method doesn't care about "improving." It cares about "clearing."

What a 13-week forecast is *not*

A 13-week forecast isn't a budget broken into weeks. Budgets are goal statements. Forecasts are probability statements. Confusing the two produces a dangerous kind of optimism: the spreadsheet says you'll have the cash because you *planned* to have the cash.

A 13-week forecast also isn't a static report. The magic is in the roll. Every week, you drop the week that just happened, add a new Week 13, and update assumptions with what reality taught you. If you don't roll it, it becomes a historical artifact wearing a future costume.

And it isn't just for companies in distress. Healthy companies use it the way good drivers use headlights: not because they expect a crash, but because they plan to see the road.

The skeleton: Beginning Balance + Receipts - Disbursements = Ending Balance

A 13-week forecast works because it's structurally simple. That simplicity is a feature: when the model is complicated, the organization stops trusting it. When the model is simple, people argue about inputs—and that's exactly where the truth is hiding.

Here is the spine you protect at all costs:

$$\text{Ending Cash} = \text{Beginning Cash} + \text{Receipts} - \text{Disbursements}$$

Everything else is detail, categorization, and governance.

How to build one that doesn't lie to you

Start with one bank account: the operating account where cash actually moves. If you have multiple accounts (payroll, sweep, savings), you can forecast them too, but don't begin by building a cathedral. Begin by building a thermometer.

Then build weekly columns: Week 1 through Week 13. Week 1 is the week you're in now (not the week you wish you were in). If you operate internationally, define "week" consistently (Monday–Sunday, for example) and make cutoffs explicit. Ambiguity is where forecast accuracy goes to die.

Now you need three input streams:

- *Beginning balance:* pull it from the bank, not the general ledger. Cash forecasting is an exercise in humility: the bank is the source of truth.
- *Receipts:* customer payments, card settlements, marketplace payouts, grants, rebates, interest—anything that increases cash.
- *Disbursements:* payroll, taxes, rent, debt service, vendors, software, capex, legal—anything that decreases cash.

The direct method is not "forecast revenue." It's "forecast collection." A deal closed in Week 3 doesn't matter if the cash arrives in Week 11.

That means your forecast must be anchored in *payment mechanics*: invoicing dates, contractual terms, historical behavior, and the ugly truth about what gets disputed.

The template that makes the method obvious

Near the point where most teams get lost—how to lay out the spreadsheet so it stays usable—the structure itself should do some of the governance for you. The model should force the conversation toward ending cash, because ending cash is the number that decides whether you can keep making choices.

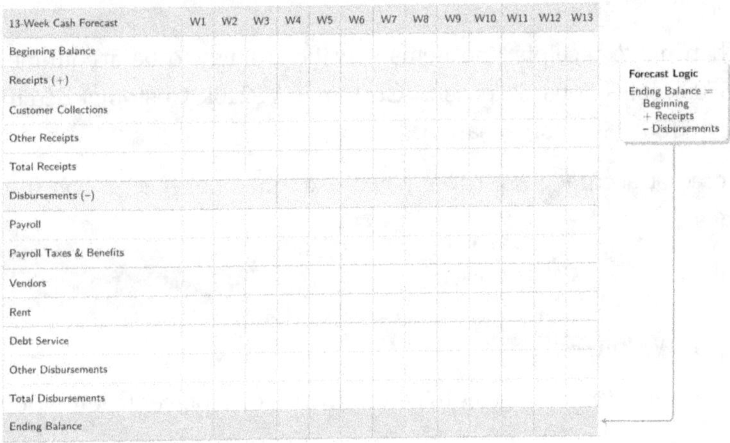

The point of the layout is not aesthetics; it's behavior. When the ending balance row is visually privileged, it becomes socially privileged. People start asking about *ending cash*, which is the only cash number that matters.

Receipts: forecast what you can collect, not what you can invoice

Receipts forecasting is where ego goes to get corrected. Sales teams live in a world of wins and pipeline. Collections live in a world of terms and friction. The forecast should live in the collections world.

Start by separating receipts into categories that behave differently:

- *Contracted recurring receipts* (subscriptions, retainers, maintenance): often predictable, especially with autopay.
- *Invoiced B2B receipts* (net terms): predictable only if you model customer behavior, not just invoice dates.
- *Card/processor settlements* (daily/weekly): predictable but can be subject to holds, chargebacks, and reserves.

- *One-time receipts* (implementation fees, milestone payments, grants): high impact, low reliability unless the trigger is already satisfied.

Then make receipts line items specific enough to be accountable. "$500k collections" is a wish. "Customer A $240k, Customer B $110k, rest $150k" is a conversation.

Forecast accuracy improves when each material receipt has three tags:

- *Amount*
- *Week expected*
- *Confidence level* (high/medium/low, or a percentage if you're disciplined)

Confidence levels feel squishy until you realize they force honesty. A low-confidence receipt should not be what prevents your ending cash from going negative. If it is, you don't have a forecast; you have a prayer.

The collections calendar: where optimism goes to die (and then gets useful)

Here's what improves receipts forecasting faster than any clever spreadsheet technique: build a collections calendar that sales, finance, and customer success can all see. Not a shame board—a scoreboard.

Put each expected customer payment on a weekly timeline with an owner. If a payment slides, it stays visible; it doesn't quietly move from Week 4 to Week 6 with no narrative. The narrative is the signal: dispute? missing PO? wrong vendor setup? AP contact changed? customer's own cash issue? Those reasons are operational data, and they help you prevent repeats.

A CFO doesn't just want cash this week. A CFO wants to reduce the future variance of cash. Variance is what turns runway into roulette.

Disbursements: the bills you can't dodge, and the ones you shouldn't

Disbursements are often easier to forecast than receipts because your obligations are within your control. They are also where organizations tell the truth about priorities.

At minimum, break disbursements into:

- *Payroll* (including commissions, bonuses when relevant)
- *Taxes* (payroll taxes, sales tax/VAT remittances, income tax estimates if applicable)
- *AP / vendors* (operational suppliers, contractors, professional services)
- *Rent and facilities*
- *Debt service* (principal, interest, fees)
- *Capex* (equipment, build-outs)

Then add two columns of nuance in your own notes (not necessarily in the sheet): *committed vs. discretionary*, and *timing flexibility*.

This is where ethics enters the forecast, whether you want it to or not. Paying slower is a lever, but it's also a relationship decision. Stretching a massive software vendor is not the same as stretching a small supplier who covers payroll by waiting for your check. A good cash forecast doesn't just show what you *can* do; it clarifies what you *should* do.

The hidden monster: payroll timing

Payroll is the most emotionally charged line item on the cash forecast, and it should be. It's the line item with names attached.

It's also the line item that frequently catches companies off guard because payroll is not just salary:

- employer taxes,
- benefits funding,
- retirement matches,
- commissions (which can be lumpy),
- bonus payouts (which are often timed to be maximally inconvenient for cash).

Put payroll in the week it will *clear*. Not the pay period. Not the date your provider drafts. The date it hits your bank account. That sounds pedantic until you miss a bank cutoff and discover that one day of timing can become one week of consequences.

The discipline: reconcile weekly, then learn from the error

A forecast that isn't compared to actuals becomes performance art. The most valuable weekly ritual takes 30 minutes:

1. Pull actual cash activity from the bank for the week that ended.
2. Compare actual receipts and disbursements to forecast.
3. Explain the largest variances in plain language.
4. Update future weeks based on what you learned.

The goal is not to be "right"; the goal is to get *less wrong in the same way*. If Customer C always pays two weeks late, your forecast should stop being surprised. If a vendor consistently drafts earlier than promised, fix the assumption or fix the relationship.

Forecasting is a competence you build through scar tissue. Weekly reconciliation turns scars into skill.

Make it operational: attach actions to ending cash bands

The ending balance row is not a trivia fact. It's a trigger system.

A practical way to make the forecast matter is to define ending cash bands and pre-decide the behaviors that come with them. For example:

- *Green band:* above your internal buffer (you can fund planned growth, pay vendors normally, consider opportunistic moves).
- *Yellow band:* above Cash Zero but below buffer (tighten approvals, accelerate collections, delay discretionary spend).
- *Red band:* approaching Cash Zero (activate the cash conservation plan, negotiate payment terms, consider financing actions, escalate cadence).

This is not about fear; it's about speed. When the band turns red, you don't want to start debating whether red is a real color.

A CFO's favorite question: ``What would have to be true?"

When someone presents a comforting 13-week forecast, treat it like a magic trick and ask to see the hands.

"We'll be fine in Week 6" is not a statement; it's a bundle of assumptions. Ask: *What would have to be true for Week 6 to look like that?* Then write those assumptions down:

- Customer A pays on time.
- Customer B's dispute resolves this week.
- The renewal closes by Friday.
- The tax payment is accurately estimated.
- The bank line remains available.

Now you have a management agenda. Each assumption can be owned, checked, and de-risked. Without this step, the forecast is just a pretty way of being surprised later.

Common failure modes (and the fixes that actually work)

Most 13-week forecasts fail in predictable ways:

- *They forecast revenue, not cash.* Fix: build receipts from customer-level expected payment dates.
- *They forget the ugly weeks.* Fix: explicitly schedule taxes, annual renewals, insurance premiums, bonus cycles, and debt service.
- *They're too detailed too soon.* Fix: start with the top 15–25 cash drivers; add detail only where it changes decisions.
- *They're owned by finance alone.* Fix: make Sales/CS own collections inputs; make Operations own major vendor outflows; keep finance as the integrator and referee.
- *They're not rolled.* Fix: make updating it part of a weekly operating cadence, like payroll itself.

The best fix is the least glamorous: consistency. A mediocre forecast updated weekly will outperform a brilliant forecast updated monthly, because reality moves faster than pride.

Actionable takeaway: build a working 13-week forecast in one afternoon

Do not start by perfecting categories. Start by making it real.

1. *Pull the bank balance today* (operating account) and make it Week 1 Beginning Balance.
2. *List the next four payroll runs, rent, debt service, and tax dates* and place them in the exact week they clear.

3. *Add the top 10 expected customer receipts* with expected payment weeks and a confidence tag.
4. *Compute Ending Balance for Weeks 1–13* using the simple spine: Beginning + Receipts – Disbursements.
5. *Circle the lowest Ending Balance* across the 13 weeks. That is your near-term risk point. Assign two actions that can move it within 14 days (collections push, spend deferral, vendor term negotiation, financing prep).
6. *Schedule a 30-minute weekly cash meeting* to reconcile forecast vs. actuals and roll forward one week. No slides. One spreadsheet. Decisions only.

If you do only one thing: protect the integrity of the ending balance row. When that row becomes trustworthy, the company starts making decisions with its eyes open—and the bank account stops being a source of surprise and becomes what it should have been all along: a managed system.

3.3 Treasury Operations: The Plumbing That Prevents Leaks

Money is the only corporate asset that can disappear in under a minute and still be "properly approved." A wire can be released at 10:02 a.m., irretrievable by lunch, and politely explained in an audit trail by dinner. That's the uncomfortable reality behind treasury work: it's not glamorous, but it is existential.

Treasury operations are the control room for cash—the place where visibility becomes action and action becomes irreversible. Forecasting tells you *when* you'll be tight; treasury tells you whether you can move cash safely without creating a new problem while solving the old one. When treasury is weak, cash management becomes a horror movie: everyone can see the threat, nobody can stop the transfer.

A good treasury function feels boring. Payments go out on time. Bank accounts reconcile cleanly. Approvals follow a routine. Nothing dramatic happens. That boredom is earned, not accidental. It is engineered.

The SLY trade: Security, Liquidity, Yield

Every CFO eventually hears the same question: "Can't our cash earn more?" The question is reasonable. It's also a trap when asked without context.

Treasury decisions live inside a three-way trade-off often summarized as *Security, Liquidity, Yield* (SLY). You can't maximize all three simultaneously. If a product offers unusually high yield with perfect liquidity and no risk, what it actually offers is a lesson in fraud.

- *Security* means you get the money back.
- *Liquidity* means you can get it back *on the day you need it.*
- *Yield* means you earn a return for letting someone else use it.

The ordering matters. Businesses fail because they lack security and liquidity, not because they lacked yield. Treasury is the discipline of keeping that ordering intact even when the market is whispering sweet nothings about basis points.

A practical rule: the closer the cash is to payroll, the less you are allowed to care about yield.

Bank accounts as architecture: separate rooms, fewer fires

Many organizations start with one bank account because it's easy, then keep it because nobody owns the redesign. One account becomes the kitchen, bedroom, and workshop of the company. That works until it doesn't.

A clean banking structure creates separation between what *must* happen and what *may* happen. It reduces accidental overdrafts, limits fraud blast radius, and makes forecasting cleaner because payment types stop stepping on each other's toes.

A common, effective structure:

- *Operating account:* primary inflows and vendor outflows. This is where you want daily visibility.
- *Payroll account:* funded just-in-time for payroll runs. Keep only what you need for the next cycle plus a cushion.
- *Tax account (optional but powerful)*: a holding pen for sales tax/-VAT and payroll tax amounts so you can't accidentally spend money that was never yours.
- *Savings / liquidity reserve:* the rainy-day buffer and near-term reserves, ideally harder to move quickly than the operating account.

The point isn't complexity; it's containment. If operating credentials are compromised, a separate payroll account can keep you from making the worst call of your career: telling employees payroll is delayed because you were robbed.

The daily habit that separates pros from improvisers: reconciliation

A cash forecast without reconciliation is a weather report written by someone who never looks out the window.

Reconciliation is the practice of matching what the bank says happened to what the company believes it authorized. Treasury teams treat this as a daily discipline because the time value of fraud is vicious. The first 24 hours after an unauthorized transfer are often the difference between "recoverable" and "we learned a very expensive lesson."

Daily reconciliation doesn't have to be ornate. It has to be fast:

- Pull bank activity (prior day, and same-day where possible).

- Match it to expected payments and deposits.
- Flag unknown items immediately.
- Escalate anything suspicious while the window for recall still exists.

This routine also catches non-fraud issues that quietly bleed cash: duplicate payments, subscription creep, bank fees that drift upward, vendor drafts that occur earlier than expected. Treasury is leak detection, and the smallest leaks compound.

The modern threat model: the ``polite email" that steals millions

Most payment fraud isn't a masked hacker pounding a keyboard in a dark room. It's a calm, well-timed email sent at 4:47 p.m. on a Friday.

Business Email Compromise (BEC) and related email/account compromise schemes exploit trust, urgency, and routine. A message looks like it's from a CEO, a vendor, a lawyer, or a project manager. The wording is plausible. The request is simple: "We updated our bank details," or "Can you send the wire today?" The amounts are often chosen to feel normal, not suspicious.

The truly dangerous part is that the payment request often conforms to your process. The invoice looks right. The approver recognizes the vendor name. The email thread is convincing. Fraud succeeds not by breaking controls, but by slipping through the gaps between them: when verification is informal, when exceptions are common, when speed is rewarded and skepticism is punished.

Treasury operations exist to make "polite fraud" hard.

Segregation of duties: the Maker-Checker principle

The oldest control in treasury is also the most effective: no single person should be able to create, approve, and release a payment end-to-end.

Call it segregation of duties, call it dual control, call it the maker-checker model—it's the same idea: separate *initiation* from *approval* from *release*.

Why it works:

- It reduces internal fraud by removing unilateral power.
- It catches honest mistakes (wrong amount, wrong vendor, wrong date) before they become expensive.
- It forces the organization to slow down just enough to notice weirdness.

A common failure mode is performing maker-checker theatrically: two approvals that happen in the same Slack thread, based on the same compromised email, by two people who both assume the other validated details. Real controls require independent verification, not shared confidence.

Design the workflow around irreversibility

Different payment rails have different "oops" profiles.

- *Wires* are fast and often irreversible once settled. Treat them like cash in an envelope.
- *ACH* is generally safer but increasingly targeted, and return windows and bank processes vary.
- *Checks* are slow, operationally painful, and historically vulnerable to alteration and theft.
- *Cards* have strong consumer protections; business card protections vary, but chargeback mechanisms create different risk dynamics.

Treasury controls should be strongest where irreversibility is highest. If a $500k wire can be released with the same friction as a $500 SaaS invoice, you don't have a control system—you have a hope system.

A practical approach is tiering:

- *Low-risk, low-dollar recurring* (e.g., utilities): streamlined approval, monitored via reconciliation.
- *Medium-risk* (standard vendors): normal maker-checker and vendor master controls.
- *High-risk* (new vendors, changed bank details, urgent wires, international payments): enhanced verification, call-backs, and senior approval.

The art is building tiers that reflect risk, not politics. "Who asked" is not a risk factor. "Can we recover the money if wrong" is.

Vendor master data: where fraud quietly sets up camp

Many fraud events don't happen at the moment of payment. They happen weeks earlier, when someone changes vendor bank instructions.

Vendor master data—the file that stores payee names, bank accounts, routing details, and payment preferences—is a high-value target. It deserves controls that feel disproportionate until you price the downside.

Minimum controls that actually work:

- *Restrict edit permissions* to a small group.
- *Separate edit from approval* (one person proposes the change, another approves).
- *Require out-of-band verification* for bank detail changes (call a known number, not the number in the email).
- *Log and review changes* weekly (a simple report: what changed, who changed it, who approved it).

Out-of-band verification is the muscle here. It feels old-fashioned. It is also one of the few defenses that reliably defeats email spoofing.

When in doubt, verify using a channel the attacker doesn't control: a phone call to a pre-existing contact, a vendor portal with strong authentication, or a confirmed contract addendum.

Positive Pay, ACH filters, and other bank-side guardrails

The bank is not just a place where money sits; it can also be part of your control system.

Depending on your bank and payment mix, treasury can deploy tools that reduce fraud exposure:

- *Check Positive Pay:* the bank matches presented checks against a file of checks you issued (number, amount, date, payee depending on setup). Exceptions require your approval.
- *ACH blocks/filters:* allow only specific counterparties or transaction types to debit your account.
- *Dual approval in the bank portal:* separate people must approve the release, not just in your ERP.
- *Transaction limits and alerts:* thresholds that trigger additional authentication or immediate notifications.

A subtle point: controls must live where the money moves. Approvals in an internal system are helpful, but if the bank portal can release funds without the same separation and thresholds, your strongest controls become optional.

Treasury teams periodically test this by simulating a payment release path: "If my laptop were compromised, how far could an attacker get?" The goal isn't to be dramatic. It's to identify the shortest path to disaster and then lengthen it.

Security is a people problem wearing a finance badge

Fraud prevention is often framed as a technology problem, but treasury leaders learn quickly that it is mostly human:

- People want to be helpful.
- People fear slowing down a senior executive.
- People interpret urgency as importance.

That's exactly why the best treasury controls are social contracts, not just system settings. A CFO can explicitly grant permission to be "difficult":

> If someone pressures you to bypass verification, you are allowed to say no—even if that someone is me.

That sentence is a control. It changes behavior. It makes fraud harder because fraud relies on shame and speed.

Training helps, but training alone fades. What sticks is ritual: the mandatory call-back on bank detail changes, the two-person release, the daily reconciliation, the weekly review of vendor changes. People remember what the organization does repeatedly.

Liquidity management meets fraud: the cruel timing of a bad week

Fraud often hits when liquidity is tight, because tight liquidity creates exactly the conditions attackers exploit: rushed approvals, stressed staff, and leadership distracted by bigger fires.

That's why treasury and cash forecasting belong together operationally. When the 13-week forecast shows a low cash week, treasury should raise the control posture, not loosen it. Tight weeks are when you feel tempted to skip steps. Tight weeks are also when you cannot afford to lose $150k to a fraudulent ACH or a misdirected wire.

A mature organization treats low-cash periods like storm warnings: increase monitoring, confirm payment details, slow down large releases, and ensure bank contacts are up to date in case a recall is needed.

Incident response: what to do in the first hour

Even with good controls, assume something will eventually go wrong. A mis-keyed account number. A compromised mailbox. A fraudulent request that slipped through. The difference between a scare and a catastrophe is how quickly you respond.

The first-hour priorities are practical:

- *Contact the bank immediately* to attempt recall/hold and to flag suspicious activity.
- *Preserve evidence* (emails, headers, payment confirmation screens, logs) without contaminating it.
- *Lock down access* (reset credentials, disable compromised accounts, review MFA).
- *Notify internal stakeholders* (finance leadership, IT/security, legal) so actions are coordinated.

Treasury teams often keep a short, printed "fraud response card" with bank contact numbers and escalation paths. In a real incident, nobody wants to hunt through shared drives for the right phone number while money is in motion.

Treasury KPIs that actually mean something

Treasury metrics are often either too high-level ("cash balance") or too abstract ("number of bank accounts"). Useful KPIs tie directly to control quality and operational cadence:

- *Bank reconciliation lag:* days between activity and reconciliation completion.
- *Percent of payments with proper dual approval in-bank:* not just in the ERP.
- *Vendor master changes per week + exception rate:* and how many were independently verified.

- *Forecast-to-actual cash variance (weekly):* not as a scorecard, but as a learning tool that improves timing assumptions.
- *Number of urgent/off-cycle payments:* urgency is where controls get bypassed; track it like a risk indicator.

These are not vanity metrics. They are early warning lights for control decay.

Actionable takeaway: a 10-point treasury hardening sprint (done in two weeks)

If treasury operations feel "fine," run this sprint anyway. Most fraud stories begin with "We didn't think it would happen to us."

1. Inventory every bank portal user; remove anyone who doesn't need access.
2. Enforce multi-factor authentication for all banking logins (and ensure it's not SMS-only if avoidable).
3. Turn on dual approval *in the bank portal* for wires and ACH, not just inside your accounting system.
4. Set sensible per-transaction and daily limits by payment type; require senior approval above thresholds.
5. Separate the operating and payroll accounts if they're commingled; fund payroll just-in-time.
6. Lock down vendor master edits to a small group; require maker-checker for any bank detail change.
7. Implement out-of-band verification for new vendors and bank changes (call a known number).
8. Enable bank alerts for large transactions, new payees, and login anomalies; route alerts to at least two people.
9. Move reconciliation to daily (or at minimum, next-business-day) for operating accounts.

10. Create a one-page incident response sheet with bank recall contacts, internal escalation, and the first-hour checklist.

If that feels like a lot, remember what treasury is: plumbing. Nobody admires it when it works. Everyone notices it when it fails, usually at the worst possible time, with the most expensive consequences. The job is not to be admired. The job is to keep the cash from leaking out of the building.

3.4 Unlocking Trapped Cash: The Cash Conversion Cycle

A profitable company can feel broke for one simple reason: it already spent the cash.

Not wasted it. Spent it *exactly* the way the business model requires—on inventory sitting in a warehouse, on payroll for work already performed, on ads that produced orders that won't be paid for weeks. The money didn't vanish. It changed shape. It became product in transit, invoices in a customer's AP queue, and promises you can't deposit at the bank.

That shape-shifting is why working capital management is so addictive once you see it clearly. Unlike price increases or new products, you can often free up cash without selling one extra unit. No heroics. Just time compression: getting cash in sooner, pushing cash out later (ethically), and keeping less of it frozen in stuff.

That time compression has a name: the Cash Conversion Cycle.

The cash conversion cycle: a time machine made of operations

The Cash Conversion Cycle (CCC) measures how long cash is tied up in the operating loop. In plain language: how many days pass between paying for what you sell and getting paid for what you sold.

A useful mental model is a relay race where cash is the baton. The baton starts as money in your bank account. You hand it to suppliers (cash out). It turns into inventory. Inventory turns into sales. Sales turn into accounts receivable (IOUs). Finally, receivables turn back into cash (cash in). The longer the baton takes to complete the lap, the more cash you need just to keep running.

Here's the classic formula:

$$\text{CCC} = \text{DIO} + \text{DSO} - \text{DPO}$$

Where:

- *DIO* (Days Inventory Outstanding): how long inventory sits before it's sold.
- *DSO* (Days Sales Outstanding): how long it takes to collect after the sale.
- *DPO* (Days Payables Outstanding): how long you take to pay suppliers.

The subtraction is the key insight: payables can partially fund inventory and receivables. Many businesses are effectively financed by their suppliers and customers—sometimes in healthy ways, sometimes in ways that strain relationships and eventually snap.

The timeline that makes the ``cash trap" visible

The formula is neat. The lived experience is messier. A timeline makes it visceral: the gap between cash leaving and cash returning is the trap you're trying to shrink.

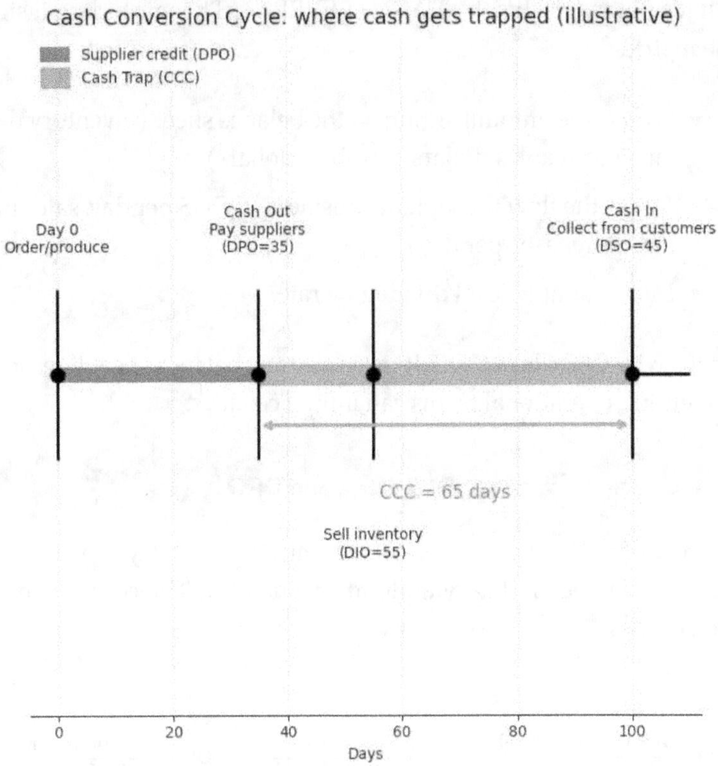

That orange bar is the part executives feel in their chest. It's the period where you've already paid (or committed to pay) for the work, but the customer hasn't paid you yet. If the bar is long, growth consumes cash even when the business is "doing well." If the bar is short, the business can fund itself more easily and withstand surprises with less drama.

CCC is not a finance trick; it's the operating model

The cash conversion cycle is derived from operational balance sheet accounts, not from the cash flow statement in a way that's easy to eyeball. That's why CCC improvements often feel like magic to non-finance teams: you don't change demand, and yet cash appears.

But it's not magic. It's the TIME = LEVEL/RATE logic hiding inside the metrics:

- *Level* is the amount sitting on the balance sheet (inventory dollars, receivables dollars, payables dollars).
- *Rate* is the flow through the business (COGS per day, sales per day, purchases per day).
- *Days* is simply level divided by rate.

That's why CFOs love CCC. It turns abstract balance sheet lines into something operational teams can influence: days.

A quick way to calculate DIO, DSO, and DPO

There are multiple correct ways to compute each "days" metric depending on your industry and data availability. The core is consistent:

$$\text{DSO} \approx \frac{\text{Accounts Receivable}}{\text{Revenue per day}} \qquad \text{DIO} \approx \frac{\text{Inventory}}{\text{COGS per day}} \qquad \text{DPO} \approx \frac{\text{Accounts Payable}}{\text{COGS per day}}$$

Using per-day rates (Revenue/365, COGS/365) keeps the math intuitive. Some teams use 360 days; the difference rarely matters compared to the operational variance you're trying to fix.

What *does* matter is consistency: use the same method over time so trends are real, not artifacts of changing math.

The emotional truth about working capital: it's where functions collide

Working capital improvement is cross-functional by nature, which is why it's both powerful and politically tricky.

- Sales wants to close. Discounts and generous terms help.

- Finance wants to collect. Tight terms and disciplined invoicing help.
- Operations wants resilience. Extra inventory helps.
- Procurement wants continuity and price concessions. Paying fast can help.

CCC is where these preferences collide in the same spreadsheet. That friction is not a sign of dysfunction; it's a sign you've found the levers that matter. The goal isn't to "win" for finance. The goal is to make trade-offs explicit and intentional, then capture cash without breaking the machine.

Lever 1: Reduce DSO (collect faster without acting like a loan shark)

DSO is the easiest lever to understand and the hardest lever to improve sustainably, because it lives inside customer relationships. Getting paid faster is partly process and partly posture.

The fastest DSO improvements usually come from fixing unsexy mechanics:

- *Invoice immediately.* "We invoice at month-end" is often code for "we like lending customers money for free."
- *Invoice correctly.* Bad PO numbers, wrong bill-to entities, and missing documentation create disputes that masquerade as "slow pay."
- *Make payment easy.* More payment methods, clearer remittance instructions, and fewer clicks matter more than your team thinks.
- *Collect on a schedule.* Collections is a calendar discipline, not an emotional improvisation.

Then come the structural moves, where money is real and feelings can get bruised:

- *Tighten terms for new deals.* It's much easier to be disciplined at the start than to renegotiate later.
- *Use deposits or milestone billing.* If you deliver value over time, bill over time. Waiting to bill until the end is a self-imposed cash drought.
- *Charge for riskier terms.* Net-90 is a financing product. If you must offer it, price it like one (higher price, smaller discount, or explicit financing fee where commercially acceptable).

A practical CFO tactic: build a "DSO bridge" that separates delays caused by *customer choice* (they pay late) from delays caused by *your process* (you invoiced late, invoiced wrong, or failed to follow up). The second category is a gift—it's within your control and doesn't require a tense customer call.

Lever 2: Reduce DIO (inventory is safety until it becomes a confession)

Inventory is where operational fear becomes expensive.

After years of supply chain disruption, many companies adopted "just-in-case" thinking: keep extra stock because lead times are unpredictable and stockouts are humiliating. That strategy can be rational. It can also quietly turn into a habit, and habits don't carry the same rigor as decisions.

DIO improves when you treat inventory like a portfolio with carrying costs, not a storage problem.

Tactics that actually move the needle:

- *Segment inventory.* Fast movers and slow movers should not be managed with the same rules. The slowest 10–20% often holds a shocking share of dollars.

- *Attack "obsolete" with urgency.* Obsolete inventory is not inventory. It's cash you already lost, sitting on a shelf to preserve everyone's feelings.
- *Improve demand planning, even crudely.* A simple forecast that is consistently updated beats an elaborate forecast that nobody believes.
- *Shorten production cycles and batch sizes.* Smaller batches can reduce work-in-process and finished goods without killing service levels.
- *Align incentives.* If operations is rewarded for never stocking out and not penalized for overstock, you will overstock. Incentives are inventory policy in disguise.

Inventory reductions should be surgical, not reckless. Slashing inventory can create stockouts that reduce revenue, which is the opposite of freeing cash. The art is to reduce *unproductive* inventory—the stuff that doesn't protect service levels, doesn't support profitable demand, and doesn't move.

Lever 3: Increase DPO (pay slower without becoming the villain)

DPO is the most misunderstood lever because it's the only one that can be "improved" instantly by simply not paying. That kind of improvement is fake. It's a short-term liquidity move that can trigger supply disruption, price increases, lost priority, and reputational damage.

Ethical DPO improvement is about redesigning terms, not breaking them.

Ways to extend payables responsibly:

- *Negotiate terms upfront.* Larger suppliers expect it. Smaller suppliers might accept it if you offer predictability (set payment days, clean remittance, fewer disputes).

- *Standardize payment runs.* Paying on set days (e.g., weekly) can extend average DPO without violating terms, because payments cluster rather than dribble out.
- *Use supply chain finance thoughtfully.* In some cases, a bank or platform can pay your supplier early while you pay later, preserving supplier health. It's not free; it's a financing decision.
- *Eliminate early-pay outflows you didn't intend.* If your AP process accidentally pays invoices as soon as they're approved, you are donating working capital.

A sharp CFO also watches for a common self-own: teams negotiate a 2/10 net 30 discount (2% discount if paid in 10 days), then *miss* the discount and still pay early. That's like ordering the expensive wine and not drinking it.

The working capital flywheel: small day improvements become big dollars

CCC improvements look small in days and enormous in cash.

A one-day reduction in DSO on $100M of annual revenue is roughly:

$$\text{Cash freed} \approx \frac{\$100M}{365} \approx \$274k$$

A one-day reduction in DIO on $60M of annual COGS is about:

$$\frac{\$60M}{365} \approx \$164k$$

A five-day improvement across the cycle can fund a meaningful hire plan, reduce revolver draws, or provide breathing room for a product reset. Working capital is one of the few places where operational tweaks can finance strategy without asking permission from capital markets.

Governance: why CCC improvements evaporate without an owner

Working capital is slippery. You can improve DSO for two months and watch it drift back as sales reintroduces generous terms to hit targets. You can purge inventory and watch it rebuild as planners re-add safety stock "just to be safe." You can extend DPO and then lose it when AP cleans up processes and inadvertently pays faster.

Sustained improvement requires governance that feels lightweight but relentless:

- A clear owner for each lever (AR, inventory, AP) with cross-functional authority.
- A weekly or biweekly cadence that reviews leading indicators (aging, past-due, inventory turns, slow-moving lists, upcoming vendor cliffs).
- Policies that hard-code decisions (standard terms, approval requirements for exceptions, required documentation for invoicing).

The tone matters. Working capital governance shouldn't feel like finance scolding people for being human. It should feel like the company deciding, together, that cash trapped in process is a waste—and that waste crowds out better uses.

A concrete example: the ``no-new-sales" cash improvement

Imagine a mid-market manufacturer with:

- \$80M annual revenue (mostly invoiced B2B)
- \$50M annual COGS
- DSO = 62 days
- DIO = 78 days
- DPO = 35 days

CCC = 78 + 62 - 35 = 105 days.

Now assume a realistic improvement program over six months:

- Reduce DSO from 62 to 52 by invoicing faster, fixing disputes, and tightening terms on renewals.
- Reduce DIO from 78 to 68 by segmenting SKUs and liquidating slow movers.
- Increase DPO from 35 to 42 by renegotiating terms with top suppliers and standardizing payment runs.

New CCC = 68 + 52 - 42 = 78 days.

That's a 27-day reduction. In rough dollars, using a blended daily operating cost base (you can proxy with COGS and operating expenses), the freed cash can easily land in the millions. The company didn't sell more. It simply stopped financing its own inefficiencies at 0% interest.

The CFO's warning label: do not optimize CCC blindly

Like any good metric, CCC can be abused.

A company can "improve" DPO by paying late and alienating suppliers. It can "improve" DIO by starving inventory and creating stockouts that destroy customer trust. It can "improve" DSO by refusing reasonable terms and losing deals.

The right objective isn't the lowest CCC. It's the best CCC *for your model*—one that supports service levels, protects relationships, and keeps the business resilient without tying up unnecessary cash.

Think of CCC optimization the way you think of dieting: losing weight by amputating a limb is technically effective and strategically insane.

Actionable takeaway: a 30-day trapped-cash hunt with real targets

Working capital improvements become real when they are specific and time-bound. Run this 30-day hunt:

1. *Compute DSO, DIO, and DPO* using a consistent method. Write down the CCC and pick a target improvement of 5–10 days.
2. *Pick one lever to lead with* based on where the cash is largest and the fixes are within reach.
3. *Launch a "top-20 list" for each balance:*
    - Top 20 past-due invoices by dollars (and why they're past due).
    - Top 20 slow-moving SKUs by dollars (and a disposition plan).
    - Top 20 suppliers by spend (and current terms vs. desired terms).
4. *Assign owners and dates* for each top-20 item. No owner, no improvement.
5. *Measure cash impact weekly* using the simple conversion: one day of DSO $\approx$ Revenue/365; one day of DIO or DPO $\approx$ COGS/365.

If the business needs cash, CCC is often the fastest place to find it because the cash is already yours. It's simply trapped in time. The job is to shorten the gap—carefully, ethically, and with enough operational respect that the improvements stick.

Chapter 4

Steering the Ship: Planning and Performance Management

If accounting is the rearview mirror, showing you where you've been, then planning is the windshield. Most companies spend too much time polishing the mirror and not enough time cleaning the windshield. This chapter transforms the finance function from a scorekeeper into a navigator. We will move beyond the dreaded annual budget exercise-often a negotiation game that ends in a spreadsheet everyone ignores-to build a dynamic system that anticipates storms, adjusts to changing winds, and actually helps the CEO make decisions before the cash runs out.

4.1 The Annual Operating Plan: Setting the North Star

The annual budget gets mocked because it's *always* wrong. A competitor cuts price in March. A key hire quits in May. A supplier changes terms in July. By September, the spreadsheet is a museum exhibit—beautifully labeled artifacts from a world that no longer exists.

And yet: companies that abandon the Annual Operating Plan (AOP) entirely tend to replace it with something worse. They don't become agile; they become *argument-driven*. Every leadership meeting turns into a fresh negotiation. Every investment request becomes a one-off debate with no shared frame. The organization can move fast, sure—but it moves fast in *different directions*.

The AOP earns its keep not by predicting the future, but by aligning the present. It is the one moment each year when the board, the CEO, and the department heads all make the same hard trade-off at the same time: *given what we know today, what are we choosing to do—and what are we choosing not to do?* That choice won't survive first contact with reality. The alignment can.

Stop Calling It a Budget and Start Calling It a Handshake

A budget is often treated like a number to be defended. An AOP is better treated like a handshake: a clear, mutual understanding of priorities, constraints, and expectations.

Handshakes have two important features. First, they're public. Everyone sees who agreed to what. Second, they acknowledge uncertainty. You can shake hands on an intention even when the details will evolve. The handshake isn't a guarantee; it's a commitment to act in good faith under a shared set of assumptions.

That distinction changes the emotional temperature of planning. When the AOP is framed as a contract, people negotiate like lawyers: protect your downside, hide your optionality, and leave room to maneuver. When it's framed as a handshake, people negotiate like operators: surface the real bottlenecks, call the risks out loud, and trade resources for outcomes.

You can feel the difference in the first fifteen minutes of a kick-off meeting. In contract-mode, the room is guarded. In handshake-mode, it's candid.

The Only Two Outcomes That Matter

Most planning processes produce two artifacts: a slide deck for the board and a spreadsheet for finance. The AOP should produce two outcomes instead.

Outcome one: a shared story. Not a slogan. A causal story. What has to be true for the company to win this year? Where will growth come from? What will we stop doing to free capacity? What risks are worth taking, and which are not?

Outcome two: a resource map. Not a line-item wish list. A map that shows where the company is placing its scarce chips: headcount, cash, management attention, and time. Time is the sneakiest constraint because it doesn't appear on the P&L. If the AOP doesn't explicitly allocate leadership attention, you've built a fantasy.

Numbers matter because they are a forcing function. But the plan isn't the numbers. The plan is the *trade-offs the numbers reveal.*

Why ``Budgets Are Useless" Is the Wrong Complaint

"Budgets are useless" is usually shorthand for three more precise complaints:

- *They take forever.* The process consumes weeks of high-value time.
- *They become obsolete quickly.* The assumptions age out fast.
- *They get weaponized.* Targets become a stick, and teams respond by gaming.

All three are solvable, but not by quitting the AOP. They're solved by changing what the AOP is *for*.

If the AOP is a prediction exercise, it fails. If it is a negotiation of priorities and capacity, it succeeds even when assumptions move. A pilot can't control the weather. The pilot still files a flight plan, because coordination matters.

The CFO's Real Job in the AOP: Make Trade-offs Explicit

CFOs often get positioned as the person who "keeps everyone honest." That's fine, but it's incomplete. The higher-value role is to make trade-offs *explicit and comparable.*

Every department can justify its own spend. Marketing can show pipeline math. Engineering can show roadmap dependencies. Sales can show quota coverage. Customer success can show churn risk. Each argument is locally coherent. Together, they don't add up.

The CFO's contribution is to translate local logic into shared decision language:

- If we hire these ten roles, what metric moves and when?
- If we delay this product investment by two quarters, what revenue shifts from this year to next?
- If we cut vendor spend, what operational risk rises?
- If we push for profitability this year, what growth do we forgo?

In other words: finance doesn't "say no." Finance makes the "yes" choices visible—and makes the "no" choices unavoidable.

The Planning Paradox: Precision Creates Fiction

AOP spreadsheets are often built with a level of precision that implies a level of knowledge no one has. Revenue to the dollar. Headcount to the decimal. Travel per person per month. It looks rigorous. It's often fiction.

Precision is not the same as accuracy. Precision is how many digits you type. Accuracy is whether the future behaves the way you assumed.

The paradox is that more precision can produce less truth. When a model looks exact, people stop interrogating the assumptions. When a model is simpler, people are forced to discuss what actually drives the outcome.

The fix isn't to be sloppy. The fix is to be honest about what is knowable at annual cadence (direction, priorities, capacity, major bets) and what belongs in shorter cycles (timing, pacing, tactical adjustments).

The V-Model: Where Strategy Meets Reality

The best AOPs are not purely top-down ("here's the number, go make it happen") and not purely bottom-up ("here's everyone's wish list, let's see what we can afford"). They are a controlled collision between the two.

Top-down brings *strategy and capital*:

- What does the board expect in growth, margin, and cash?
- What strategic moves are non-negotiable?
- What is the company's risk appetite this year?

Bottom-up brings *resources and tactics*:

- What does it actually take to deliver the roadmap and the pipeline?
- Where are the bottlenecks (hiring market, onboarding capacity, tooling, product constraints)?
- What is already committed (contracts, leases, minimum staffing)?

The AOP is the meeting point: a plan that is ambitious enough to matter and realistic enough to execute.

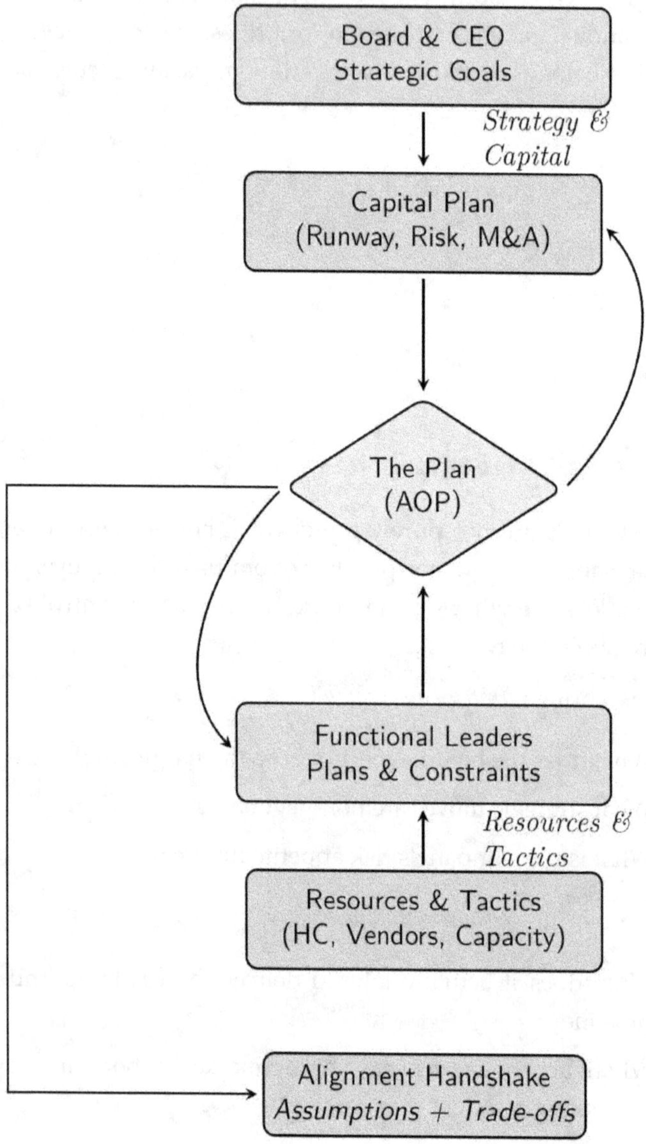
Board & CEO
Strategic Goals
Strategy &
Capital
Capital Plan
(Runway, Risk, M&A)
The Plan
(AOP)
Functional Leaders
Plans & Constraints
Resources &
Tactics
Resources & Tactics
(HC, Vendors, Capacity)
Alignment Handshake
Assumptions + Trade-offs

What Alignment Actually Looks Like (When It's Working)

Alignment isn't everyone agreeing. Alignment is everyone disagreeing *in the same place*, with the same facts, and then committing to the decision.

When the AOP is healthy, you'll see specific behaviors:

- Sales and marketing argue about pipeline math, not about who "owns revenue."
- Engineering pushes back on timelines with evidence about throughput and dependencies, not with vague claims of complexity.
- The CEO makes at least one visible trade: "We're delaying initiative B to fund initiative A."
- The board conversation shifts from "why is expense up?" to "what are we buying with expense?"

If none of those moments happen, you likely produced a budget but not an AOP. The plan was never really negotiated; it was merely assembled.

The AOP Is Not a Forecast (And That's a Feature)

This is where a lot of organizations tangle themselves: they try to make one document serve three purposes.

- *A target* (what we want to achieve)
- *A forecast* (what we realistically think will happen)
- *A resource allocation* (what we are willing to fund)

Those can't peacefully coexist in one number set, because they produce different incentives. Targets pull you upward. Forecasts pull you toward truth. Resource allocations force trade-offs.

A mature finance function separates them conceptually, even if leadership prefers a single headline. The AOP is primarily a *resource allocation and priorities* document anchored to targets. The forecast is the living truth-teller that updates as reality changes. When you confuse the two, teams learn that "forecasting" is a performance evaluation ritual, and candor evaporates.

The AOP can still be anchored by a top-line and margin goal. But the operating truth of the company should not be held hostage to the political needs of target-setting.

How Gaming Sneaks In (And How to Block It)

Gaming isn't a moral failure. It's a natural response to a system.

If hitting the budget is how people keep their jobs, budgets will be negotiated downward and padded upward. Managers will delay expenses to make a quarter look good, then binge the next quarter. Sales teams will sandbag pipeline early and "miraculously" close late. Engineering will under-promise to preserve credibility. Everyone becomes rationally defensive.

Two practical moves reduce gaming dramatically:

First: separate "commit" from "aspire." A simple structure is:

- *Commit*: what the business can achieve with high confidence if the plan executes.
- *Aspire*: what becomes possible with upside conditions (faster sales cycles, stronger conversion, earlier launches).

People can be held accountable to the commit while still being motivated by the aspire.

Second: fund outcomes, not departments. This doesn't mean reorganizing the company. It means forcing each major budget request to answer a single question: *what outcome does this buy, and what is the evidence chain from spend to result?* When that becomes the norm, padding looks embarrassing because it lacks a story.

AOP Mechanics That Keep You Out of Spreadsheet Hell

The fastest way to kill the AOP is to turn it into an annual census of every expense code.

AOP detail should be proportional to controllability and materiality:

- *Headcount*: detailed by role and timing because it's expensive, lumpy, and operationally constrained.
- *COGS and gross margin*: detailed by major drivers (unit cost, hosting, fulfillment, mix), not by every invoice.
- *Opex*: detailed where decisions exist (new programs, big vendor renewals), rolled up where it doesn't (small tools, minor travel).
- *One-time items*: explicitly called out so they don't silently become "run-rate."

A useful rule: if changing a line by 10% would not change any decision, it doesn't deserve an hour of debate.

AOP as a Capacity Plan: The Hidden Constraint Is People-Time

Finance naturally thinks in dollars. Execution happens in hours.

Two plans can have identical P&Ls and radically different chances of success depending on management bandwidth. A company can "afford" six initiatives and still fail because it can only *run* three at a time.

Treat the AOP as a capacity plan by explicitly naming:

- The top initiatives that will consume cross-functional time (launches, system migrations, pricing changes).

- The bottleneck teams (data engineering, security, legal, recruiting, onboarding).
- The calendar realities (holiday seasons, peak demand periods, renewal cycles).

When those are named, the plan becomes executable. When they're ignored, the plan becomes a wish that turns into a post-mortem.

A Practical Way to Run the Negotiation (Without Breaking Relationships)

Most AOP conflicts are not about the math. They're about mismatched narratives:

- The CEO hears "we can grow 30%."
- Sales hears "we need 12 reps."
- Finance hears "we can afford 8 reps."
- Recruiting hears "you won't get 12 qualified reps in time anyway."

The negotiation becomes productive when you force the conversation into the same units:

- *Capacity*: how many deals can the team realistically handle?
- *Timing*: when do hires become productive?
- *Conversion*: what will actually move win rates or cycle time?
- *Cash*: what can the company survive if the ramp is late?

Once those are on the table, the "budget fight" turns into an operating design problem. That's a fight worth having.

A Mini Case: The Plan That Failed for a Non-Financial Reason

A mid-market B2B software company built an AOP that looked immaculate: 25% growth, modest headcount expansion, improving margins. The math worked. The board approved. Everyone exhaled.

By Q2, the plan was off-track. Not because demand collapsed, and not because costs blew up. The plan failed because implementation work swallowed the year.

The company had committed to a CRM migration, a billing-system overhaul, and a new product launch in the same two quarters—each of which required the same scarce group of people: sales ops, RevOps, product ops, and engineering leads. The AOP had funded the work but hadn't allocated *attention*. Everyone was "fully resourced" on paper. In reality, the same humans were booked three times.

The lesson isn't "don't do big projects." The lesson is that an AOP that only balances dollars is incomplete. A plan is a promise about throughput, and throughput is made of people-time.

What the Board Actually Wants From the AOP

Boards rarely want a perfect forecast. They want a coherent stance.

A good AOP gives a board confidence in three areas:

- *Economic clarity*: what drives revenue, margin, and cash in this business.
- *Strategic intent*: what management is choosing to prioritize and why.
- *Control*: how management will know early if the plan is breaking, and what levers exist.

When you present the AOP as a stance rather than a prophecy, the tone of board governance improves. The board stops asking for certainty you can't provide and starts asking for signals and options you *can* provide.

AOP Outputs That Are Actually Useful in March

The AOP should not die on January 2nd. To keep it alive, produce outputs designed to survive contact with reality:

- *A one-page plan*: the handful of commitments and trade-offs everyone can remember.
- *A short list of assumptions*: the few assumptions that, if wrong, change decisions (pricing, churn, hiring velocity, conversion).
- *A "stop-doing" list*: what will be deprioritized if capacity tightens.
- *Guardrails*: cash minimums, hiring triggers, spend approval thresholds tied to performance.

Notice what's missing: a 40-tab workbook as the centerpiece. The workbook still exists. It just isn't the product.

Actionable Takeaway

Write down three sentences and force leadership to agree on them before anyone debates line items:

1. *This year we are optimizing for* ________________ *(growth, profitability, cash, or a specific strategic move).*
2. *We will fund* ______________________ *even if it hurts near-term optics, because it's our bet.*
3. *If the world changes, the first thing we will stop or slow is* ______________________.

Then build the AOP so the numbers *prove* those sentences are true. If the model contradicts them, you don't have a budget problem—you have an honesty problem.

4.2 Dynamic Planning: Replacing Crystal Balls with Drivers

A finance team can spend six weeks building an annual plan and still get blindsided by a single Tuesday.

A big customer pauses renewals because their CFO froze spending. A new product tier quietly cannibalizes the old one. A viral moment drives traffic up 60% and support tickets up 120%. None of those events show up as a neat line in an annual spreadsheet. They show up as a mess: a spike in volume, a shift in mix, a hiring scramble, a cash wobble.

Dynamic planning is the move from *numbers-as-fate* to *numbers-as-sensors*. Instead of asking, "What will revenue be in November?" you build a model that answers, "If pipeline coverage changes, if conversion slips, if churn ticks up, what happens next—and when do we feel it in cash?" The goal isn't clairvoyance. The goal is early warning and fast coordination.

Forecasts Are Flashlights, Not Report Cards

The fastest way to destroy a forecasting system is to turn it into a performance evaluation ritual. The moment managers believe their credibility, compensation, or headcount depends on the forecast being "optimistic enough," the forecast becomes theater. People stop reporting what they see and start reporting what they think leadership wants to hear.

A rolling forecast works when it's treated as a flashlight: it illuminates the path ahead so the company can avoid stepping into holes. Flashlights don't get graded for being "right." They get used.

That means two rules that feel almost too simple:

- The forecast is allowed to be bad news.
- The forecast is required to be specific about why it changed.

Bad news with no causal explanation is panic. Bad news with drivers is management.

The Rolling Forecast: A Horizon That Moves with You

A rolling forecast keeps a constant forward-looking horizon—often 12–18 months. When a month closes, you don't just fill in actuals and call it done. You *add* a new month at the end and refresh the periods in between.

This sounds like a small procedural tweak. It's not. A fixed annual budget turns time into a cliff: everything beyond December becomes someone else's problem. A rolling forecast turns time into a runway: you can see the next several quarters continuously, which is where hiring plans, cash decisions, and capacity constraints actually live.

A practical example: imagine it's February. A fixed plan might care intensely about March and barely acknowledge next January. A rolling forecast cares about March *and* next January because both will be in view, and both will be influenced by decisions you make this week (especially hiring, pricing, and multi-quarter projects).

Reactive vs Proactive: The Same Tool, Two Very Different Cultures

Rolling forecasts tend to fall into two camps.

Reactive rolling forecasts are essentially "budget re-forecasts." They exist to update the outlook after the month closes. The conversation is backward-looking: what moved, how far off are we, can we still hit the annual number.

Proactive rolling forecasts are run like an operating system. They are updated frequently enough, and with enough driver-level clarity, that leaders use them to trigger actions early: slow hiring, accelerate a product launch, shift marketing mix, renegotiate a vendor contract, adjust pricing, change payment terms, tighten collections.

The difference is not sophistication. It's intent. Reactive forecasting treats the forecast as a *scoreboard*. Proactive forecasting treats it as a *control panel.*

If you want the proactive version, the forecast needs three design features:

- It updates on a predictable cadence (monthly is common; some businesses do key drivers weekly).
- It has a short list of drivers that senior leaders actually understand.
- It connects to decisions with triggers ("if X happens, we do Y").

Driver-Based Modeling: Stop Forecasting Dollars, Start Forecasting Reality

A line-item budget asks, "How much will we spend on travel?" That's like asking, "How much will we eat this year?" It's a downstream consequence, not a controllable input.

Driver-based modeling flips the model so you forecast what the business *does*, and let dollars fall out of the math.

A driver is an operational variable that causes financial outcomes. A few common ones:

- B2B SaaS: leads, conversion rate, average selling price (ASP), churn, expansion, implementation capacity, support tickets per customer
- E-commerce: sessions, conversion rate, average order value (AOV), returns rate, fulfillment cost per order, ad CAC
- Marketplace: supply hours, demand requests, take rate, match rate, refund rate
- Services: billable hours, utilization, realization, average rate, project duration

Drivers have two magical properties:

- They are closer to truth. Your VP of Sales may not know next quarter's revenue, but they can tell you pipeline, cycle time, and win rates. Your head of Support can't predict total opex, but they can predict ticket volume and handle time.
- They create shared language. Operations and finance can argue productively about "conversion fell 80 bps" instead of trading feelings about "marketing spend is too high."

The point is not to model everything. The point is to model *the few things that move the business.*

A Good Driver Is Both Predictive and Actionable

Finance teams often pick drivers that are easy to measure but not useful to manage. Others pick drivers that are "strategic" but too vague to forecast. A good driver sits in the narrow middle: it helps you see around corners *and* gives you a lever to pull.

Here's a quick sniff test:

- If the driver changes, does it materially change a financial outcome within the horizon?
- Can a leader plausibly influence the driver with a real action (pricing, staffing, process change, product changes, marketing mix)?
- Does the business have a reliable way to measure it without heroic manual work?

If the answer to any of these is "no," you don't have a driver. You have a dashboard ornament.

The Model You Actually Need (Not the Model That Impresses Analysts)

Many teams build driver models that resemble elaborate clocks: dozens of interlocking gears, all exquisitely labeled, none of which anyone can repair when it breaks. The model becomes a priesthood. Only FP&A can touch it. Everyone else stops trusting it because it's opaque.

A model earns trust when it is:

- *Small*: it fits on a whiteboard. Not literally, but emotionally.
- *Transparent*: a leader can trace "if conversion drops, revenue drops" without needing a finance translator.
- *Stable*: the structure doesn't change every month; only the assumptions do.

A useful mental template is "one page of drivers, one page of outputs." Drivers are the knobs. Outputs are the P&L, cash, and key operational capacity constraints.

From Drivers to Dollars: The Chain Matters

Driver-based planning fails when the driver is picked correctly but the translation to dollars is sloppy.

Take support tickets. If tickets rise 30%, costs don't necessarily rise 30%. They rise based on capacity, productivity, and service levels:

- tickets per agent per day
- target response time
- percent handled by automation
- escalation rate to engineering (a hidden cost multiplier)

This is where finance can be surprisingly creative. You're not just forecasting spend; you're modeling the operating design. If the company wants to keep customer experience flat while volume rises, you can quantify what that promise costs and when it hits.

The best models make it obvious where the business has leverage. Sometimes the cheapest "cost reduction" is not cutting headcount; it's reducing rework, improving onboarding, or fixing a product defect that's generating tickets like a machine.

The Most Common Failure Mode: Confusing the Forecast with the Target

Targets are motivational. Forecasts are descriptive.

If leadership says, "We *need* to hit $50M ARR," that's a target. If FP&A says, "Based on pipeline, conversion, and churn, we're likely to land at $46M," that's a forecast. When targets and forecasts are forced to be the same, one of them becomes dishonest.

A healthy cadence allows both to exist:

- The target stays steady enough to align effort.
- The forecast moves as reality moves.
- The gap between them becomes a management problem to solve, not a reporting problem to hide.

When you hear someone say, "Update the forecast to show we'll hit the number," you're no longer forecasting. You're negotiating.

A Cadence That Feels Like Management, Not Bureaucracy

Dynamic planning is not a quarterly ritual where everyone panics, updates spreadsheets, and returns to ignoring the model.

A sustainable cadence has three tiers:

- *Weekly (lightweight):* track a handful of leading drivers (pipeline creation, conversion, churn signals, usage, ticket volume, cash collections). This can be a dashboard, not a forecast refresh.

- *Monthly (serious)*: refresh the rolling forecast with the latest actuals and updated driver assumptions. Lock a version, not because it's "right," but because it's the basis for decisions.
- *Quarterly (strategic)*: revisit structural assumptions (pricing changes, product strategy, hiring plan shape, market shifts) and run scenarios.

The monthly cycle is where finance earns trust: not with fancy math, but with consistent follow-through. People start to believe: "If we tell finance what's actually happening, the model will reflect it, and leadership will act on it."

The CFO's Secret Weapon: Triggers and Guardrails

Driver-based rolling forecasts become powerful when they are tied to pre-agreed actions. Otherwise the company simply watches the numbers deteriorate with increasing sophistication.

Examples of triggers and guardrails:

- If forecasted cash dips below a minimum buffer, freeze nonessential hiring and require CFO approval for new vendor commitments.
- If pipeline coverage falls below a threshold for two consecutive weeks, shift spend from brand to demand gen, or redeploy sales development capacity.
- If churn in a key segment rises above a set level, pause expansion hiring and fund retention initiatives.
- If gross margin compresses due to mix, tighten discounting rules or renegotiate key COGS inputs.

Notice the psychological advantage: you are not improvising under stress. You are executing a plan you agreed to while calm.

A Short, Concrete Example: From ``Sales Is Behind" to a Driver Diagnosis

Imagine a company misses revenue in the first month of the quarter. In a traditional budget world, the story quickly becomes moral: "sales execution is weak," "marketing leads are low quality," "pricing is wrong," "the product isn't ready." The argument heats up because there's no shared diagnostic.

A driver-based forecast forces a different conversation:

- Pipeline created is on plan, but win rate is down 3 points.
- Average selling price is down 5% due to discounting in one segment.
- Sales cycle length has increased by 12 days, pushing closes out of quarter.

Now the question isn't "who failed," it's "which lever matters most and what can we do *now*?"

- Tighten discount authority and reposition a bundle.
- Add sales engineering capacity to reduce cycle time.
- Focus reps on late-stage deals and adjust qualification.

The forecast becomes a map of interventions, not a postmortem.

Designing the Driver Set: Fewer Than You Think

There's a temptation to model the entire business because it feels responsible. The result is a model that requires a small army to maintain and produces little clarity.

Most companies can get enormous value from 6–12 primary drivers, grouped into a few themes:

- Demand: pipeline creation, conversion, price/discount
- Retention: churn, expansion, usage health

- Capacity: headcount, ramp time, productivity
- Economics: unit cost, margin by product/segment, key input inflation
- Cash mechanics: collections timing, payment terms, deferred revenue dynamics (if applicable)

The discipline is to pick drivers that are *owned* by leaders. If a driver has no owner, it will be updated late, argued endlessly, and quietly ignored.

Linking to Cash: Where Dynamic Planning Becomes Real

P&L forecasting is emotionally charged, but cash forecasting is where companies live or die. Dynamic planning becomes materially better when drivers flow through to cash implications:

- If sales slips, when do billings slip?
- If billings slip, when do collections slip?
- If collections slip, does cash fall below the buffer before you can react?
- If you slow hiring, when does that actually reduce cash burn (consider severance, recruiting costs already incurred, timing of start dates)?

Cash introduces time delays. Those delays are exactly why a rolling horizon matters. A decision made in February can determine whether you're forced into a painful capital raise in October.

Implementation Reality: Keep Consolidation Separate from Modeling

A quiet killer of dynamic planning is overloading the forecast model with reporting detail. Consolidation needs GAAP-aligned accounts, entities, cost centers, and elimination logic. Forecasting needs speed, drivers, and flexibility.

If you try to build one system that does both perfectly, you often end up with a forecast that takes too long to update and a consolidation process that's constantly being "tuned."

A practical compromise:

- Keep the driver model at a level where leaders can engage (business unit, product line, major cost pools).
- Map driver outputs into the chart of accounts for reporting, but don't make leaders forecast at GL-line granularity.

The forecast is a decision tool. The close is a compliance and credibility tool. They can talk to each other without being the same thing.

Actionable Takeaway

Build a one-page driver forecast in a week, even if it's imperfect:

1. Pick 8–10 drivers with clear owners (e.g., pipeline created, win rate, ASP/discount, churn, expansion, headcount, ramp time, productivity, unit cost, collections timing).
2. Write the translation rules in plain language (e.g., "Revenue = deals × ASP," "Deals = pipeline × win rate," "Cash in = billings collected with 45-day lag").
3. Set two triggers tied to cash and one trigger tied to growth (e.g., cash buffer minimum, hiring threshold, pipeline coverage threshold) and pre-agree what action follows.

Don't wait for the "perfect" model. A simple driver forecast that gets updated every month will outperform a beautiful spreadsheet that only gets touched when leadership is already worried.

4.3 Driving Accountability: The Monthly Business Review

The monthly close is when finance learns what happened. The Monthly Business Review (MBR) is when the company decides what to do about it.

That sounds obvious until you watch a leadership team treat the MBR like a courtroom. Finance plays prosecutor. Department heads play defense. Someone waves a spreadsheet like an alibi. The meeting ends with vague instructions ("tighten spend," "push harder," "be strategic") and a calendar invite for next month's trial.

A good MBR feels different. It feels like a pit crew. The car is still moving. Everyone is looking at the same gauges. The questions are blunt, not personal: what changed, why did it change, and what are we changing *now* so next month's numbers don't tell the same sad story with new formatting.

Accountability Isn't Blame. It's Memory with Consequences

Companies without an MBR don't lack data. They lack *institutional memory*. The same surprises keep happening, and each time they get explained as an "anomaly." An MBR is a mechanism that says: if something moves, we name it; if we name it, we decide; if we decide, we track whether we did it.

Accountability, in this sense, isn't about catching someone. It's about building a loop tight enough that reality can't be ignored for long. The best leadership teams don't fear that loop. They rely on it, because it replaces hallway gossip with shared facts.

Two Modes: Diagnostic and Interactive (You Need Both)

Every MBR has two jobs, and confusing them ruins the meeting.

Diagnostic mode is the classic: actuals versus plan, variance versus budget or forecast, and a clear view of what's off and by how much. Diagnostic reviews are supposed to be boring in a healthy company. Boring means the machine is stable and the exceptions are visible.

Interactive mode is different. It's where leaders use a small set of measures to debate strategic uncertainties: pricing pressure, a changing sales cycle, capacity constraints, competitive moves, a shift in customer behavior. Interactive mode is where a forecast becomes a decision tool instead of a score.

A common failure is running diagnostic mode for 90 minutes and interactive mode for zero. The meeting becomes a forensic report delivered after the accident. The company gets very good at explaining bruises.

A better pattern is to timebox diagnostic review and then deliberately switch gears: "Given what we're seeing, what do we believe has changed in the business? What are we doing about it before it hits cash?"

The MBR Is a Product. Finance Is the Product Manager

Treating the MBR as a calendar event is like treating your pricing page as a PDF. You can do it, but it's a choice to be ignored.

A great MBR has product qualities:

- *A clear user*: the executive team (and anyone who owns a number in the plan).
- *A clear purpose*: decisions and commitments, not narration.
- *A consistent interface*: the deck looks and feels familiar every month so attention goes to what changed, not to learning the format again.

Finance owns the interface, the definitions, and the discipline. Not because finance is power-hungry, but because someone must protect comparability over time. If every month reinvents the metrics, you can't tell improvement from rearrangement.

Variance Analysis: The Art of Asking "Compared to What?"

"Revenue was $10.2M." That's a number. It isn't information.

Information starts with comparison:

- Versus plan (the AOP's intent)
- Versus last month (momentum)
- Versus last year (seasonality and scale)
- Versus forecast (the best current view)

The trick is to avoid turning the MBR into a comparison buffet. Too many baselines create ambiguity: any result can be made to look good or bad depending on which comparison you pick. Choose one primary baseline for accountability (often the latest approved forecast) and keep the others as context, not escape hatches.

Then comes the more important move: *decomposition*. A variance is only useful when you can break it into causes that are operationally meaningful.

For revenue, the most useful decomposition is often Price–Volume–Mix.

Price, Volume, Mix: Three Words That End Unproductive Arguments

Revenue misses trigger moralizing. "Sales isn't executing." "Marketing leads are weak." "The product is overpriced." These are claims you can't settle in a room without turning it into a debate club.

Price–Volume–Mix (PVM) variance analysis forces specificity:

- *Price variance*: did we sell at a different price (or discount) than planned?

- *Volume variance*: did we sell more or fewer units (or subscriptions, seats, transactions) than planned?
- *Mix variance*: did the composition of what we sold change (more low-margin products, fewer enterprise deals, different customer segments)?

The magic is that none of these are "excuses." They are *mechanisms*. Mechanisms can be managed.

Price variance can be addressed with discount policy, packaging, competitive positioning, deal desk rigor, or value-based selling. Volume variance is usually about demand generation, sales capacity, conversion, or churn. Mix variance often reveals strategy drift: the business is winning, but winning in a less profitable or less scalable way than intended.

A Waterfall Makes the Story Impossible to Dodge

Executives are busy. Your job is not to give them more data. It's to make the truth hard to avoid.

A waterfall chart (sometimes called a bridge) does exactly that. It turns a fuzzy "we missed revenue" into a walk from Budget to Actual, with each driver taking its share of the blame or credit. It's visual accountability: the bars are the argument.

Place it next to the explanation, and the meeting changes. People stop arguing about whether the miss is "real." They start arguing about what to do about the specific bars.

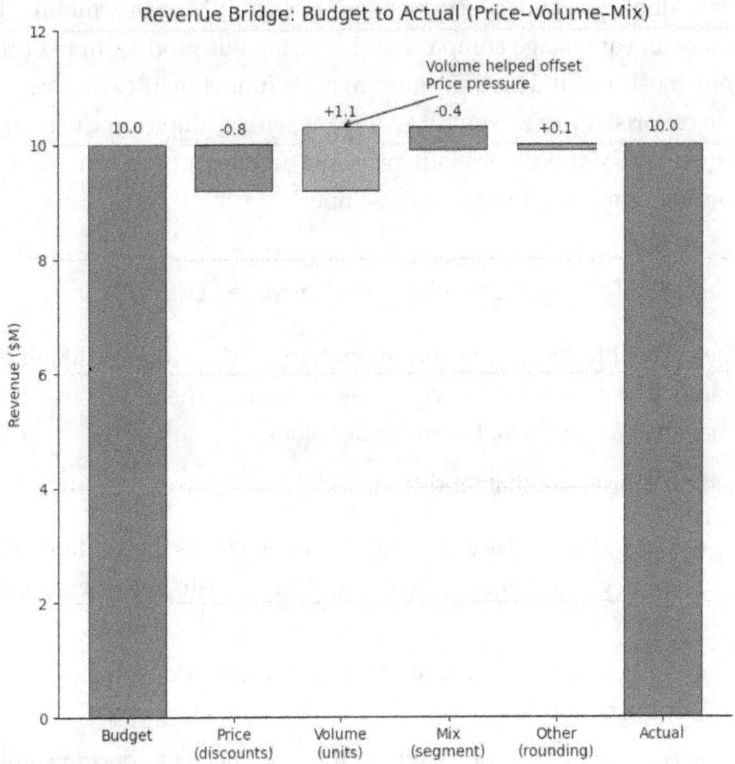

How to Run PVM Without Turning It into a Math Fight

PVM can become its own rabbit hole if people argue about definitions ("Is that mix or price?") instead of actions. The goal is not perfect attribution. The goal is a shared diagnosis good enough to decide.

Two practical guardrails keep it productive:

- *Agree on a consistent method and stick to it.* Changing the method month to month is a stealth way to avoid accountability.
- *Tie each variance to an owner and an action.* A variance without an owner is trivia. A variance without an action is lament.

Also: don't pretend every business needs PVM every month. If you're an early-stage company still figuring out product-market fit, your most useful decomposition may be funnel metrics and retention cohorts. If you're a mature company with multiple products and segments, PVM becomes indispensable because mix can quietly eat your margin while revenue looks "fine."

The MBR Deck That Doesn't Waste Everyone's Life

The best MBR decks are short enough to finish and rich enough to matter. The goal is not to "cover the business." The goal is to cover what changed and what requires decisions.

A simple structure that tends to work:

- *Scoreboard (one page):* the handful of outcomes that matter (revenue, gross margin, operating income, cash, key operational metrics).
- *Bridges:* revenue bridge (PVM), margin bridge (price, cost, mix), cash bridge (working capital drivers if relevant).
- *Focus areas:* 2–4 topics where the team needs to decide something (capacity, hiring, pricing, churn, collections).
- *Actions and owners:* a tight list of commitments with dates.

If you can't fit actions and owners on one page, the meeting is producing heat, not movement.

The Quiet Power Move: Leading Indicators Next to Lagging Ones

Financial results are lagging indicators. They tell you what happened after it already happened.

A well-run MBR pairs lagging outcomes with leading indicators that predict the next 60–120 days. Not because it's trendy, but because it changes behavior. People stop optimizing for the accounting period and start optimizing for the business.

Examples that tend to matter:

- Revenue next quarter is often more sensitive to pipeline creation and cycle time than to this month's bookings.
- Margin next quarter is often sensitive to discounting discipline and product mix, not just vendor invoices.
- Cash next quarter is often sensitive to collections and payment terms, not just "spend control."

This is where a Balanced Scorecard mindset becomes practical rather than academic: you don't want twenty metrics; you want the few operational signals that explain and predict the financial outcomes.

Corrective Action: The Meeting Isn't Over Until Something Changes

A depressing number of MBRs end with a vibe. The vibe might be concern, determination, or mild shame. None of those pay the bills.

A corrective action is something you can track. It has:

- a specific change (do X, stop Y, decide Z),
- an owner,
- a deadline,
- and a metric that will move if the action worked.

If a revenue miss is driven by price variance, "sell harder" is not a corrective action. "Reinstate approval thresholds for discounts above 20% and require deal desk review for nonstandard terms, effective immediately" is an action. If volume is the issue, "increase pipeline" is not an action; "add two SDRs to enterprise accounts, reallocate $50k from brand to high-intent search, and review daily meeting set rates" is an action.

The CFO's role is to be politely relentless here. Not mean. Not cynical. Just unwilling to let the company buy another month of denial.

Common Traps (That Smart Teams Still Fall Into)

Even strong teams fall into predictable traps:

The "numbers readout" trap. Someone narrates every line item. The deck becomes an audiobook. People tune out and then ask questions that were answered three slides ago.

The "variance confessional" trap. Leaders explain misses with long stories but no quantification and no decision. The meeting becomes therapeutic, not operational.

The "everything is a one-time item" trap. One-time items happen. So do recurring patterns disguised as one-time items. The MBR should be where patterns get named.

The "finance owns the truth" trap. Finance should own definitions and consolidation. Leaders should own the drivers and the actions. If finance is the only function that can explain the numbers, the business is flying without instruments.

A Human Detail That Changes Everything: Psychological Safety

This is the part that doesn't show up in templates.

Teams only share early warning signs when it's safe to do so. If the MBR punishes candor, the business learns to delay reality. Leaders wait until problems are undeniable. By then, the options are worse and the cost is higher.

Paradoxically, a demanding MBR can be the *most* psychologically safe meeting in the company if the norms are clear:

- We challenge numbers, not people.
- We reward early flags, even when they're ugly.
- We separate explanation from action: you can explain without defending, and you can commit without posturing.

This is why "gotcha" finance cultures eventually get blindsided. They train the organization to hide the smoke until the building is on fire.

Actionable Takeaway

At the next MBR, replace every "why" question with a decomposition question, then force an action:

1. For the biggest revenue variance, require a Price–Volume–Mix bridge (even if it's rough). Don't accept "execution" as an answer.
2. For the biggest expense variance, separate *rate* (cost per unit) from *volume* (units consumed). A lot of "overspend" is actually "more activity," which might be good.
3. End the meeting with a single slide titled *Decisions and Commitments*, listing owners, deadlines, and the metric each action is meant to move. If the slide is empty, the meeting didn't happen.

Do that consistently for two cycles and something surprising happens: people start walking into the room already thinking in drivers, already carrying proposals, already expecting decisions. The MBR stops being a monthly performance and becomes what it was always supposed to be—a control room.

4.4 Navigating Uncertainty: Scenarios and Stress Tests

A forecast is a single line drawn through a fog bank. It's useful—until you realize the fog has terrain. Cliffs, not just hills.

Leaders tend to treat uncertainty like a nuisance. "We'll adjust when we know more." That sounds sensible. It's also how companies wake up to a cash problem with no good moves left. When the bad outcome arrives, it arrives already carrying momentum: pipeline has softened for months, customers have quietly stretched payment terms, inventory has piled up, and the hiring plan is still running because nobody wanted to be the first person to blink.

Scenario planning is the discipline of blinking early *on purpose*. It takes uncertainty out of the realm of vibes and turns it into a set of explicit stories with numbers attached. Stress testing goes one step further: it asks, without flinching, "What would it take to actually break us?" Not in a cinematic way. In a very ordinary way: a cash balance that falls below zero, a covenant trip, a payroll you can't meet, a refinancing you can't get.

The Real Enemy Isn't Uncertainty. It's Surprise

Uncertainty is normal. Surprise is optional.

Surprise happens when the organization has no shared language for "things are changing." The sales team sees cycle time stretching but doesn't want to be pessimistic. Marketing sees CAC rising but argues it's "temporary." Operations sees a supplier wobble but assumes procurement will handle it. Finance sees cash conversion slowing but gets told, "That's just timing."

Scenarios create a neutral container where people can say hard things without sounding like they're rooting against the company. You're not declaring defeat; you're rehearsing responses.

The payoff is emotional as much as financial. A leadership team that has rehearsed the downside is calmer when signals deteriorate. Calm teams make better decisions.

Base, Best, Worst: Simple Enough to Use, Serious Enough to Matter

Sophisticated scenario frameworks exist, and many companies never use them because they're too ornate. The most effective starting point is still the classic: Base, Best, Worst.

The trick is to give each case a clear identity:

- *Base case* is not "last year plus 10%." It's the most realistic view of what happens if current trends continue and the plan executes as designed.
- *Best case* is not wishful thinking. It's a coherent upside: which drivers improve, why, and what constraints might stop you from capturing it.
- *Worst case* is not "we miss by a little." It's a severe-but-plausible downside that forces you to confront your real vulnerabilities.

A useful litmus test: if the worst case doesn't make someone uncomfortable, it's probably not a worst case. It's a slightly sad base case.

Scenarios Should Be Stories, Not Spreadsheets

A spreadsheet can calculate anything. That's the problem. It happily adds 2% here and subtracts 3% there until you get a neat output that means nothing.

A scenario should be a story with internal logic:

- What changes in the world?
- How does that change customer behavior?
- How does customer behavior change your drivers?
- How do driver changes flow into P&L, balance sheet, and cash?

Make the story concrete. "Enterprise budgets freeze for two quarters" is a story. "Revenue down 15%" is a symptom. Stories create shared intuition, and intuition is what you use when the data is incomplete.

Correlation Is the Whole Game

The biggest advantage of scenarios over simple sensitivities is correlation. In real life, bad things travel in groups.

When demand weakens, pricing pressure often rises. When pricing pressure rises, discounting increases, which can attract lower-quality customers who churn faster. When churn rises, support workload rises, and sales teams get distracted by renewals. Meanwhile, collections slow because customers are managing their own cash. At the same time, lenders and investors become choosier, which raises your cost of capital or delays fundraising. None of those relationships appear if you do isolated "what if revenue drops 10%" math.

A scenario forces you to model interaction effects: the second-order consequences that turn a manageable hit into a compounding problem.

This is why the best scenario work feels less like accounting and more like operational chess. The numbers are important, but the shape of the chain reaction is the real insight.

Risk Appetite, Risk Capacity, and the CFO's Hard Sentence

Most executives can discuss risk in abstract terms. Very few can translate it into operating constraints.

Two definitions make the conversation tangible:

- *Risk capacity* is how much pain the business can absorb without breaking—the real limits set by cash, covenants, funding access, and operational resilience.
- *Risk appetite* is how much pain you are *willing* to accept in pursuit of upside.

The CFO's hard sentence is: "We can afford to be wrong by ___________, but not by ___________."

That sentence should not be delivered as drama. It should be delivered as a fact derived from stress testing: cash buffers, fixed costs, working capital dynamics, debt terms, and the speed at which you can take corrective action.

Stress Tests: The Question That Changes the Room

Scenario planning asks: "What might happen?"

Stress testing asks: "What would break us?"

Breaking can mean different things:

- Cash drops below zero (or below a minimum buffer that keeps payroll safe).
- A debt covenant is breached (leverage, fixed charge coverage, minimum liquidity).
- A refinancing window closes and the maturity wall becomes existential.
- A supplier or operational failure creates a step-function increase in cost or a revenue interruption.

The power of the stress test is that it converts abstract fear into measurable thresholds. Once you know your thresholds, you can build triggers and management actions that keep you away from the edge.

It also exposes where you've been lying to yourself. Not maliciously. Just the normal human version: assuming you can cut costs instantly, assuming customers will keep paying on time, assuming lenders will be friendly, assuming hiring can be turned off like a faucet.

Start With Cash: The Universal Constraint

You can survive a bad quarter on the P&L if you have cash and time. You cannot survive a cash crunch with a great narrative.

That's why stress tests should start with liquidity: cash on hand, revolver availability, and the timing of inflows and outflows. Then layer in the mechanisms that move cash:

- billings and collections timing
- gross margin and variable cost behavior
- payroll and fixed commitments
- working capital swings (inventory, payables, receivables)
- capex and contractual obligations

A stress test that doesn't specify *when* cash becomes tight is incomplete. Timing is the difference between "we can manage this" and "we're already out of time."

The Most Dangerous Part of a Stress Test: Management Actions

Stress tests often look reassuring because they assume management can respond perfectly and immediately. That's not planning; that's fiction with a footer.

Make management actions explicit and realistic:

- What can you cut quickly without destroying revenue generation?
- What costs are sticky (severance, contractual minimums, facilities, vendor commitments)?
- How long does it take to slow hiring in practice (interviews in flight, signed offers, ramp)?
- What actions have second-order effects (cutting marketing reduces pipeline; cutting support increases churn)?

Also model *behavioral* changes in a downturn. Customers don't just buy less; they pay later, push for discounts, and demand more proof. Suppliers may tighten terms. Employees may leave. Your DSO can creep while your DPO shrinks—a nasty combination that doesn't show up in a revenue-only scenario.

The CFO's credibility is built on not assuming heroics.

A Concrete Example: ``Revenue Down 20%" Is Not a Scenario

"Revenue down 20%" is a destination. It's not a map.

A real downside scenario might look like:

- Pipeline creation drops 25% because a channel dries up.
- Win rates fall 2 points because competitors discount aggressively.
- Sales cycles extend by 15 days as procurement scrutiny increases.
- Renewal churn increases from 8% to 11% as customers rationalize tools.
- Collections slow by 10 days because finance teams at customers are stretched.

Now you can see levers:

- shift marketing to high-intent channels
- tighten discount governance and repackage value
- add enablement to reduce cycle time
- focus customer success on retention plays
- proactively renegotiate payment terms with key customers

This is where scenarios earn their keep: they turn the future into a set of operational decisions.

Designing Scenarios Without Turning Them into Fan Fiction

A common fear is that scenarios will become storytelling contests. The antidote is to anchor each scenario in a small set of driver shifts and to require a rationale for each shift.

A practical design approach:

- Pick 4–6 drivers that account for most of the business's variability (volume, price/discount, churn, unit costs, collections timing, hiring pace).
- For each scenario, change only those drivers—but change them in a correlated way.
- Force each driver shift to have a reason: competitive pricing, macro slowdown, supply disruption, product issue, regulatory change.

If someone proposes a downside where revenue falls but churn improves and collections accelerate, ask why. Sometimes weird combinations happen. Most of the time, someone is unconsciously smoothing the story to make it less scary.

Severity and Plausibility: The Twin Constraints

Stress tests must be *severe* enough to matter and *plausible* enough to engage leadership.

If a scenario is too mild, it's a comfort blanket. If it's too extreme, leaders dismiss it as science fiction and you lose attention.

One way to calibrate plausibility is to look at:

- your own historical volatility (best and worst quarters)
- industry downturn behavior (pricing pressure, churn spikes, DSO creep)
- concentration risk (what happens if your top 3 customers delay renewals?)

- operational single points of failure (a key supplier, a single data center region, one critical team)

The goal is not to predict the next crisis. The goal is to pick scenarios that expose the company's fragility and force pre-decisions.

Second-Round Effects: The Part Everyone Forgets

The first-round effect is the direct hit: fewer units sold, lower price, higher churn.

The second-round effect is what happens because you reacted:

- A hiring freeze protects cash, but slows product delivery and causes longer-term competitive damage.
- Cutting marketing protects margin, but starves pipeline and makes the next two quarters worse.
- Aggressive discounting protects volume, but trains customers to wait for discounts and compresses future pricing power.

A high-quality stress test includes second-round effects explicitly, even if the math is rough. It forces an honest discussion about trade-offs: what are you willing to damage to survive, and what must be protected at all costs?

This is also where governance matters. In a real downturn, decisions will be made fast. If you haven't pre-aligned on what is sacred (customer trust, product reliability, key talent, minimum cash buffer), panic will choose for you.

Turning Scenarios into Triggers: When Will You Act?

Scenarios are valuable. Triggers are transformative.

A trigger is an observable signal that tells you which scenario you are drifting into. It's the difference between "we planned for this" and "we are debating whether it's happening."

Examples of triggers:

- Pipeline coverage below X for two consecutive weeks
- Net revenue retention below Y for a month
- DSO up by more than Z days
- Gross margin down by X points due to mix shift
- Cash forecast crossing a minimum liquidity threshold within N months

Tie triggers to *pre-agreed actions*. Otherwise triggers become trivia, too.

The best triggers are uncomfortable because they force action before pain is obvious in GAAP results. That's the point.

The Cultural Benefit: Permission to Say ``I Don't Know" with Structure

One of the most underrated benefits of scenario work is that it gives leaders permission to admit uncertainty without sounding incompetent.

Without scenarios, saying "I don't know" feels like weakness. With scenarios, "I don't know" becomes a structured statement: "We're seeing signals consistent with the downside case on pipeline and collections; here are the actions we should take while we learn more."

That's not weakness. That's adult leadership.

Actionable Takeaway

Run a 90-minute cash stress test with your executive team using only four inputs and one output:

1. Set a minimum cash buffer (the number below which you refuse to operate).

2. Pick a severe-but-plausible shock to revenue drivers (price, volume, churn) and a plausible collections slowdown (DSO).
3. Model the timing: when does cash cross the buffer if you do nothing for 60 days?
4. List management actions you can actually execute in 30 days, and re-run the timing with those actions included.

The output is a single sentence everyone remembers: *"In this downside, we hit the buffer in ______ months unless we do ________________ by __________."*

That sentence turns uncertainty into governance. It gives the CFO the confidence to move early—and gives the rest of the team a reason to follow.

Chapter 5

Funding the Vision: Capital Structure and Fundraising

Money acts as the fuel for your corporate engine, but not all fuel burns the same. Every dollar you bring into the business comes with a price tag-whether it's the interest on a bank loan or the slice of ownership you hand over to an equity partner. This chapter moves beyond the mechanics of simply 'getting cash' to the strategic art of 'funding wisely,' helping you design a capital structure that accelerates growth without sacrificing your control or future flexibility.

5.1 The Price of Fuel: Understanding Cost of Capital

Every company runs on fuel. Sometimes that fuel is literal—raw materials, payroll, a warehouse lease. But underneath all of it is a quieter kind: the money you borrow, retain, or raise to pay for everything else. That money is never free. Even when it arrives as "founder-friendly" equity or a "cheap" revolver, it comes with a meter running in the background: the return that your providers of capital expect you to generate for them.

A CFO's job gets dramatically easier once you treat cost of capital as a *rental rate*. You don't "own" the bank's money. You rent it. You don't "own" shareholders' patience. You rent it too—in the form of expected growth, future cash distributions, or a credible path to liquidity. When you rent anything expensive, you should know the rate *before* you sign the contract, not after you've built the factory.

Cost of capital is a price, not an opinion

Many internal debates about investment sound like debates about taste: "I feel good about this product," "This market is hot," "Competitors are doing it." Cost of capital is the antidote to vibes. It forces a simple translation: *How much return do we need, in percentage terms, to justify consuming scarce capital?*

That percentage is not chosen by management. The market sets it through interest rates, credit spreads, investor expectations, and the perceived risk of your business model. You can negotiate the edges (terms, structure, timing), but you cannot negotiate gravity. If you deploy $10 million into a project that reliably returns 6% when your blended cost of capital is 11%, the project might still "grow revenue," "expand footprint," and "build the brand." It also quietly transfers wealth away from whoever funded you. In the long run, that shows up as a lower valuation, tighter covenants, or investors who stop taking your calls.

A useful mental model: *profit is not value*. Value is profit after you pay the rent on the money you used to earn it.

Two prices: what debt costs and what equity demands

There are only two fundamental ways to fund a business: with debt or with equity. Everything else is a remix.

Debt has an explicit price: the interest rate, plus fees, plus whatever economic pain is embedded in covenants and collateral requirements. If you miss the payments, the lender has contractual remedies. Debt is usually cheaper because lenders get paid first and have legal protections. It's also less forgiving because the payment schedule does not care about your product roadmap.

Equity has an implicit price: investors don't invoice you monthly, but they expect a return commensurate with risk. Their "payment" is a combination of future dividends/buybacks (for mature companies) and future exit value (for growth companies). Equity is usually more expensive because equity holders are last in line and absorb the most uncertainty. It's also more patient in the sense that it doesn't trigger default when a quarter goes sideways—but it will punish you in other ways: dilution, governance constraints, or a lower future multiple if trust erodes.

This creates the central tension of financing: debt is cheaper but brittle; equity is expensive but flexible. Your cost of capital is the blended price you pay for the mix you choose (and the mix the market will tolerate).

WACC: the blended rental rate for your balance sheet

Weighted Average Cost of Capital (WACC) is the standard way to express that blended price. Think of it as the interest rate on the whole company—not the rate you pay the bank, but the rate the *business* must earn on its invested capital to satisfy everyone who funded it.

The classic form is:

$$\text{WACC} = \left(\frac{E}{D+E}\right) \cdot r_E + \left(\frac{D}{D+E}\right) \cdot r_D \cdot (1-T)$$

where:

- E is the market value of equity
- D is the market value of debt
- r_E is cost of equity (what equity investors require)
- r_D is cost of debt (your borrowing cost)
- T is the marginal tax rate (because interest is generally tax-deductible, creating an after-tax benefit)

The formula looks academic until you remember what it's doing: it's pricing the two pools of capital by how much of each you use. If you lean more on debt, the cheap part of the blend gets heavier. If you lean more on equity, the expensive part dominates. The tax term recognizes a real-world discount: interest typically reduces taxable income, so the *after-tax* cost of debt is lower than the headline coupon.

Two quick warnings that save careers:

1. *WACC is not constant.* It moves with interest rates, spreads, operating risk, leverage, and market sentiment.
2. *WACC is not a "company score."* A lower WACC is not automatically "better." It can be lower because you're safer—or because you're more leveraged and closer to distress. Cheap fuel is great until the engine starts knocking.

A CFO-friendly way to estimate WACC without pretending it's precise

WACC is often taught like it's measured with a micrometer. In operating life, it's measured with a wrench. You want a number that is directionally right, consistent, and easy to update when assumptions change.

A practical approach:

Step 1: Estimate your cost of debt (r_D)**.** Use the rate you would pay *today* for incremental borrowing, not the weighted average interest rate on existing debt from last year's notes. The market doesn't care what your old loans cost; it cares what new capital costs now.

If you have a term sheet, use that all-in yield: interest rate plus amortized fees. If you don't, estimate based on:

- a base rate (SOFR or a government bond yield as a proxy for "risk-free")

- plus a credit spread consistent with your leverage, coverage ratios, industry cyclicality, and collateral package

CFO intuition check: if your cash flows are volatile, your cost of debt is not just "rates are high." It's the lender pricing your probability of becoming a problem.

Step 2: Choose a tax rate (T) that matches reality. In textbooks, T is a clean corporate tax rate. In real companies, taxes behave like weather: predictable in patterns, messy in detail. Use a marginal rate that reflects the incremental tax benefit of interest deductions. If you have NOLs, valuation allowances, or operate across jurisdictions, your effective marginal rate may differ from your statutory rate.

Also be honest about limitations: tax benefits are only valuable if you can use them. The theoretical "tax shield" is not a coupon you can cash at the bank.

Step 3: Estimate cost of equity (r_E) as a required return, not a promise. Cost of equity is the hardest component because it's not directly quoted. In public markets, it's often estimated using CAPM (risk-free rate + beta × equity risk premium). In private markets, you can still use that logic, but you'll need judgment around beta, size premiums, and company-specific risk.

You don't need to worship CAPM to use it well. The real point is conceptual: equity requires a premium over "safe" returns because equity takes the first punch.

If you want a CFO-grade estimate:

- start with a long-term risk-free proxy (e.g., government bond yield)
- add an equity risk premium (what investors demand for stocks over risk-free assets)
- adjust for operating risk (cyclicality, customer concentration, pricing power)

- adjust for leverage (more debt makes equity riskier)
- add a company-specific premium when the story includes real execution risk (product-market fit, regulatory exposure, key-person dependence)

The output is not "true." It's useful. The difference matters.

Step 4: Weight debt and equity using market-value thinking. WACC weights should reflect the economic share of financing, not just accounting values. For public companies, market cap is a straightforward equity value. For private companies, you can approximate equity value using recent financings or a valuation range from your fundraising discussions. For debt, book value is often a decent proxy unless the debt is meaningfully off-market.

If your WACC changes by 200 bps depending on whether you use book or market weights, that's a signal: your capital structure is shifting under your feet, and your "normal" assumptions are stale.

Where WACC goes wrong (and how to keep it honest)

WACC is powerful, but it has sharp edges. A CFO who uses it mechanically can accidentally institutionalize bad decisions.

Mistake 1: Using a single WACC for projects with different risk. A mature maintenance capex project in a stable business line should not be judged by the same hurdle as a speculative new product in a new geography. Using one WACC for everything is like pricing all insurance policies the same, regardless of whether the driver is a cautious parent or a teenager with a sports car.

Fix: use *risk-adjusted hurdle rates*. Start with corporate WACC as a baseline, then add (or subtract) a risk premium depending on project characteristics.

Mistake 2: Treating WACC as a reason to lever up forever. The formula appears to reward debt because it is cheaper and tax-advantaged. In real life, debt is cheap until it's catastrophic. The cost of financial distress is not linear. It arrives as a cliff: lost customers, distracted leadership, vendor tightening, layoffs that damage execution, and renegotiations that hand control to creditors.

Fix: pair WACC with solvency and liquidity constraints. If a capital structure lowers WACC by 50 bps but meaningfully increases default probability in a downturn, you're not optimizing; you're gambling with a weighted coin.

Mistake 3: Using yesterday's inputs. A WACC built on last year's rates is a trap. When the cost of capital moves, your "value creation" math changes even if the project didn't. This is one reason boards sometimes feel whiplash: the same project looks great in one environment and mediocre in the next, without any change in operational plan.

Fix: update WACC on a cadence (quarterly for many businesses; monthly in volatile environments), and explicitly show which changes are driven by markets versus by company fundamentals.

Mistake 4: Confusing WACC with your actual cash cost. WACC includes the cost of equity, which is not a cash expense. That makes people dismiss it. That dismissal is expensive. Equity may not be a cash cost, but it is an *ownership cost.* If you treat equity as free because there's no coupon, you'll issue too much of it, too cheaply, too often—and wake up one day realizing you've built a great business for someone else.

Fix: treat dilution like a financing cost. When you model scenarios, include ownership outcomes as seriously as interest expense.

Hurdle rates: the internal speed limit that keeps you from value destruction

WACC tells you what your capital providers require in expectation. A *hurdle rate* is the return threshold you use internally to approve investments. In a well-run finance organization, WACC is the foundation; hurdle rates are the operating policy.

Why not just use WACC as the hurdle? Because WACC is an average. Your decision set is not average.

A hurdle rate should reflect:

- *risk* (project volatility, reversibility, competitive response)
- *option value* (does this investment unlock future choices?)
- *execution uncertainty* (can your organization actually deliver?)
- *capital rationing* (what do you have to *not* do if you do this?)

The best hurdle rates are not heroic. They are protective. They prevent a company from gradually filling its calendar and balance sheet with "good stories" that compound into mediocre economics.

A story you've seen: the expansion that looked profitable until it met the cost of capital

Picture a regional services company that wants to expand into a neighboring state. The operating plan looks clean: hire a sales lead, lease a small office, buy equipment, and ramp over 18 months. The pro forma says the new branch will generate $5 million in revenue at a 12% EBITDA margin by Year 3. Everyone nods. It "works."

Then finance asks one question: *how much capital is tied up to make that $600k of EBITDA?*

The answer is usually bigger than people think:

- upfront equipment and setup
- working capital (receivables grow before cash collections stabilize)
- management time (which has an opportunity cost even if it's not booked)

Suppose the expansion consumes $4 million of incremental invested capital by Year 3 (net of payables). EBITDA of $600k sounds nice, but EBITDA is not after-tax operating profit, and it ignores maintenance capex. After converting to an after-tax operating profit proxy, maybe you're at \$350k–\$400k of true operating earnings. On \$4 million, that's roughly a 9%–10% return on invested capital.

If your WACC is 11%, you just found a professional way to lose money while feeling productive. The branch may still be strategically necessary, but now the conversation is adult: either (1) improve economics (pricing, utilization, capex discipline, collections), (2) reduce invested capital (asset-light models, vendor financing), or (3) admit it's a strategic bet and price it with a higher hurdle and tighter milestones.

This is why cost of capital matters: it separates "busy growth" from value creation.

ROIC vs. WACC: the value-creation scoreboard

There's a simple scoreboard that turns cost of capital into daily operating relevance:

$$\text{Value creation (economic spread)} = \text{ROIC} - \text{WACC}$$

ROIC (Return on Invested Capital) measures how efficiently the business generates after-tax operating profit from the capital employed in operations. It's the operating engine's efficiency rating. WACC is the cost of renting the engine.

When ROIC exceeds WACC, you're creating value; when it's below, you're destroying value. The difference is called the *spread*. Spreads compound. A 200 bps positive spread sustained for years is how "boring" companies become great. A negative spread sustained for years is how charismatic strategies become restructuring decks.

The practical insight: a CFO doesn't only manage cash. A CFO manages *spreads*. Pricing, working capital discipline, capex rigor, product mix, and customer concentration all change ROIC. Capital structure and market conditions change WACC. Your job is to widen the gap between them without taking existential risk.

How to institutionalize hurdle rates without killing initiative

Hurdle rates fail when they become a blunt weapon: "Finance says no." Used well, they become a creative constraint: "If we want to do this, we must design it to clear the bar."

A lightweight governance model that works in practice:

- *Set a corporate baseline hurdle rate* anchored to WACC (updated on a cadence).
- *Define two to three risk tiers* with clear criteria (e.g., core maintenance, adjacent expansion, new venture). Each tier has a premium over baseline.
- *Require a spread narrative,* not just a spreadsheet: what specifically drives ROIC (pricing, utilization, retention, working capital) and what could break it.
- *Use milestones for high-uncertainty bets*: approve the next tranche of capital only when leading indicators (not lagging revenue) are achieved.

This approach respects entrepreneurial energy while protecting the balance sheet. The organization still takes bets; it just stops taking unpriced bets.

One-page CFO worksheet (the version you actually use)

If you want something that can live in a memo template or investment committee deck, this is the compact version:

1. *Corporate WACC estimate (range)*: ____________________ to ____________________
2. *Baseline hurdle rate*: ____________________
3. *Project risk tier premium*: ____________________
4. *Project hurdle rate*: ____________________
5. *Projected ROIC (Year 3 steady-state)*: ____________________

6. *Economic spread (ROIC – hurdle):* ________________________
7. *Two biggest ROIC drivers:* ____________________________
8. *Two biggest ROIC risks:* ____________________________
9. *Capital at risk before proof:* ________________________
10. *Kill/continue milestones:* ____________________________

This worksheet has a hidden superpower: it forces every champion to articulate *how* value is created, not just *what* is being built.

Actionable takeaway

Pick one active investment proposal (or one recently approved project) and run a "cost of capital reality check" in under an hour: estimate a WACC *range* (not a point), assign a hurdle rate based on risk tier, and compute the implied economic spread using a simple ROIC proxy. If the spread is negative or fragile, don't cancel the project by default—redesign it. Ask, in writing: *What must change in pricing, capital intensity, or working capital to clear the hurdle without relying on optimism?*

5.2 The Balancing Act: Designing Your Capital Structure

A capital structure is your company's load-bearing design. It's the part nobody sees in the glossy strategy deck, but it determines whether the building stands when the wind picks up. Two businesses can sell the same product, at the same margin, to the same customers—and one sleeps at night while the other lives quarter-to-quarter, negotiating with its lenders like a tenant begging for a rent extension.

The uncomfortable truth is that leverage doesn't usually kill companies in good years. It kills them in *ordinary* years: the year a big customer delays a rollout, a new competitor steals a few points of price, or receivables stretch from 45 days to 70 because someone at your customer changed their AP process. If your balance sheet can't tolerate that kind of boring disappointment, it's not *efficient*. It's fragile.

What capital structure is actually for

The simplistic view is that capital structure is a choice between "cheap debt" and "expensive equity." The operating view is more useful: capital structure is a tool for allocating three scarce resources.

- *Cash flow timing*. Debt is a schedule. Equity is a horizon. Your mix determines whether a temporary dip is a memo to the board or a crisis call to the bank.
- *Control*. Debt brings rules (covenants, reporting, collateral). Equity brings partners (board seats, vetoes, preferred rights). You're choosing which constraints you want to live with.
- *Option value*. A resilient balance sheet gives you choices: to buy a competitor, to keep hiring in a downturn, to invest when others freeze. A tight balance sheet turns every decision into a trade: "If we do this, what breaks?"

If you remember nothing else: capital structure isn't just financing. It's *operating freedom* priced in monthly payments and governance terms.

The clean theory (and why it's still useful)

In a frictionless world, capital structure wouldn't matter. That's the famous Modigliani—Miller baseline: if there are no taxes, no bankruptcy costs, no information gaps, and everyone can borrow at the same rate, the value of the firm is independent of how it's financed. It's a provocative idea because it forces a sharper question: if the financing mix doesn't create value by itself, then any real-world advantage must come from real-world mess.

And real-world mess is exactly what CFOs manage for a living:

- taxes and the limits of deductibility
- distress costs and the brutal nonlinearity of "almost default"
- information asymmetry (you know more than investors; they price that)
- agency and governance (what people do when incentives and oversight change)

So the point of the theory isn't to say "don't worry about capital structure." It's to say: don't kid yourself that a clever mix of paper can rescue a mediocre operating model. Financing is an amplifier, not a generator.

Leverage: the amplifier with a hair-trigger

Leverage is simple: you fund some assets with borrowed money, and the equity gets the residual. When business is good, the residual looks heroic. When business is merely okay, the residual shrinks. When business is bad, the residual can flip negative fast.

A CFO can respect leverage without romanticizing it by keeping two definitions in mind:

- *Financial leverage* is the presence of fixed obligations: interest, scheduled principal, leases, minimum payments.
- *Operating leverage* is fixed cost in the P&L: plants, salaried headcount, long-term contracts.

These two levers multiply each other. A company with high operating leverage is already a geared machine. Adding heavy financial leverage on top is like driving a sports car on ice: exhilarating right up to the moment it isn't.

Risk magnification: what ROE hides until it's too late

Executives love Return on Equity (ROE) because it tells a flattering story when leverage works. The problem is that ROE is a *residual metric*. It tells you what's left for equity holders after everyone else has been paid. That makes it incredibly sensitive to small changes in operating performance.

Below is the simplest way to make this visceral: same business, different balance sheet, and two economic states. In the good year, leverage looks like genius. In the bad year, it looks like a lawsuit waiting to happen.

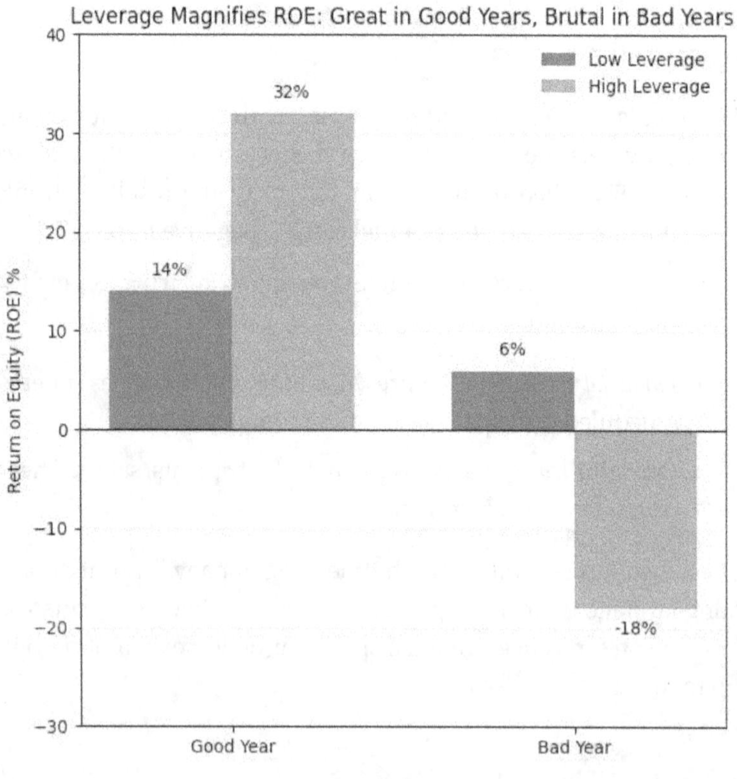

That picture is why sophisticated boards don't ask, "Can we increase ROE?" They ask, "What happens to coverage and covenants in a downturn?" A capital structure that only works in the good year is not a structure. It's a bet.

The real trade: tax shield versus distress cost

Debt often lowers the blended cost of capital because interest is generally tax-deductible. That tax shield is real—it's money you don't send to the government. But it comes with an opposing force: the expected cost of distress. Distress costs are not just legal fees in bankruptcy court. They start earlier, when the market senses weakness:

- customers demand shorter contracts or pricing concessions ("will you be around?")
- suppliers tighten terms ("prepay or no shipment")
- employees leave quietly (and take institutional memory with them)
- management spends more time on lenders than on customers

The cruel part is that distress costs spike nonlinearly. A move from 1.5x to 2.5x leverage can be manageable; a move from 4.5x to 5.5x can be fatal, depending on cyclicality and cash conversion. That's why "optimal" leverage is never a single number. It's a range that depends on how ugly your ugly case can get.

Taxes also have nuance. The tax shield is not unlimited. Interest deductibility can be constrained by tax rules (and those constraints can interact with profitability, EBITDA, and your entity structure). Treat the tax benefit as an *earned* advantage, not a permanent entitlement.

The pecking order: why equity can be more expensive than the spreadsheet says

If you've ever watched a management team avoid issuing equity even when it seems "cheaper," you've seen information asymmetry in action. Investors know you know more than they do. When a company chooses to issue equity, markets sometimes infer: "Management thinks the stock is expensive," or "They're patching a hole," or "Cash flows aren't as strong as they look." That inference can create a valuation haircut, which increases the true cost of equity right when you're trying to raise it.

This behavior has a name in finance theory (the pecking order): firms tend to prefer internal funds first, then debt, and issue equity last. Not because CFOs hate dilution on principle (though they often do), but because equity issuance can be a *signal* that investors price.

In private markets, the same dynamic shows up as tougher terms: heavier liquidation preferences, participating preferred, more control rights, or ratchets. Those are not cosmetic. They are the "interest rate" embedded in equity when the market senses that you need it.

The operating takeaway is blunt: if you wait until you *need* equity, you'll pay more for it—in valuation, in terms, and in time.

A practical decision tree: what your business can safely carry

You can't copy another company's leverage ratio the way you copy their product features. Capital structure capacity comes from business mechanics, and business mechanics are stubborn.

Here are the underwriting questions that actually matter.

How predictable is gross margin? High variability in input costs, project overruns, or pricing power means your earnings can swing. Lenders hate swings. If your gross margin moves like a seismograph, your debt capacity is lower than your last twelve months suggests.

How fast do you turn profit into cash? A business with slow collections and heavy inventory needs liquidity even when it's "profitable." The cash conversion cycle is a silent lever on debt capacity. If you have to fund 90 days of working capital, leverage is not a spreadsheet ratio; it's a daily stress test.

How reversible are your costs? Can you cut spend quickly without breaking the business? If your cost base is locked in (leases, minimum volume contracts, specialized headcount), your downside is steeper and your safe leverage is lower.

How concentrated is your revenue? If two customers represent 45% of revenue, your lenders will underwrite those customers as much as they underwrite you. Customer concentration is leverage in disguise: it magnifies volatility when relationships change.

How cyclical is demand? Cyclical businesses can carry debt, but the structure must respect the cycle: amortization that doesn't strangle you at the trough, covenants with headroom, liquidity reserves that are real (not "available if we hit the forecast").

Metrics that matter (and what they miss)

Capital structure debates often devolve into a handful of ratios. Ratios are helpful, but they can also hypnotize.

Net leverage (Net Debt / EBITDA). This is the headline number because it's compact. The danger is in the denominator: EBITDA can be flattering, cyclical, or full of "adjustments." CFO-grade leverage work uses multiple earnings views: reported EBITDA, run-rate EBITDA, and a recession-case EBITDA that you would defend under oath.

Interest coverage (EBIT or EBITDA / interest). Coverage is closer to the real question: can you pay the rent? But it ignores principal amortization and working capital swings. A company can have fine interest coverage and still suffocate on scheduled amortization and seasonal cash needs.

Fixed-charge coverage. This improves the picture by including leases and other fixed payments. It's less glamorous, more honest.

Liquidity headroom (cash + revolver availability). This is the metric people look at only when they're scared, which is exactly backwards. Liquidity is the difference between a strategic reset and a restructuring. If the next 90 days of liquidity depends on "nothing unexpected happens," you don't have liquidity. You have hope.

None of these metrics are sufficient alone. The skill is triangulation: ratios for structure, cash forecasts for timing, and scenarios for survival.

Designing for survival: the recession case is the real base case

A healthy balance sheet is not designed around your plan. It's designed around your *volatility*. The best CFOs build a recession case that is specific enough to be uncomfortable:

- revenue down X% (based on your own history or industry patterns)
- margin compression due to discounting or mix
- working capital deterioration (slower collections, inventory build, supplier tightening)
- a delayed ability to cut costs (because layoffs and renegotiations take time)

Then they ask a question that is both boring and existential: *Do we stay in compliance and liquid without heroic actions?*

"Heroic actions" include selling assets quickly, raising equity in a down market, or renegotiating covenants under pressure. Those are not plans. Those are what you do when you didn't plan.

Covenants are not handcuffs---until they are

Covenants get described as restrictions, which makes them feel like an insult. A better framing: covenants are the lender's early-warning system and your forced operating discipline. The problem is not covenants; it's covenants without headroom.

Two CFO moves separate adult debt from dangerous debt:

- *Negotiate for headroom when you don't need it.* When you *need* it, you're buying it at retail.
- *Match covenants to how your business actually behaves.* If your EBITDA is seasonal or project-based, quarterly tight tests can create "false positive" defaults at exactly the wrong time.

The goal is not covenant avoidance. The goal is to ensure that a normal wobble doesn't become a technical default that steals your negotiating power.

Structure is strategy: the hidden costs of ``cheap" debt

Even when the interest rate is attractive, structure can make debt expensive:

- *Amortization that fights your cash cycle.* If principal payments ramp before the project pays back, you've financed long-term assets with short-term urgency.
- *Security and collateral.* Pledging critical assets can limit future borrowing capacity and complicate M&A.
- *Restrictive baskets and negative covenants.* Limits on acquisitions, capex, dividends, or even certain operational moves can quietly reduce the value of opportunities.

- *Change-of-control clauses.* These can turn an attractive exit into a refinancing scramble.

A CFO who only negotiates rate is negotiating the cover price. The real economics are inside the terms.

A working philosophy for ``right-sized" leverage

There's no universal optimal mix, but there is a philosophy that travels well across industries:

> Use debt when you can predict the cash, not when you want to improve the math.

Predictable cash can come from recurring revenue, long-term contracts with strong counterparties, essential services, or dominant market position. When predictability is low, equity is often the more honest instrument because it shares risk rather than forcing a schedule onto uncertainty.

Another rule that saves time: if the main argument for more leverage is "our competitors do it," you're already losing the plot. Competitors' capital structures are often artifacts of their history, not proof of your safety. You don't inherit their mistakes; you only inherit your own.

Actionable takeaway

Write a one-page "Downturn Capacity Memo" for your current capital structure and share it with your CEO:

1. Define a specific recession case (revenue, margin, working capital) that you believe could plausibly occur.
2. Build a 13-week cash forecast and a 12-month covenant forecast under that case.

3. State, in one sentence, the earliest month you would breach a covenant or run out of liquidity *without* heroic actions.
4. If that month is uncomfortably soon, propose one structural change (more equity, longer tenor, less amortization, more headroom, a larger liquidity buffer) and quantify how much time it buys.

If you can't say when the balance sheet breaks, you don't yet have a capital structure—you have a guess with interest attached.

5.3 Going to Market: The Realities of Raising Funds

Raising money has a strange side effect: it turns sensible adults into fortune-tellers. A lender wants to know what your cash flow will look like twelve quarters from now. An equity investor wants to know whether your market will tip in your favor before your competitors learn to copy you. Everyone asks for certainty. Your job is to sell them *confidence* without selling them *fantasy*.

The fastest way to lose leverage in a fundraising process is to treat it like paperwork. Fundraising is sales, yes—but it's also underwriting. Every question you're asked is a proxy for a fear: "Will I get paid back?" "Will I get diluted into irrelevance?" "Will this team surprise me?" When you learn to hear the fear underneath the question, you stop answering defensively and start designing the process.

Two markets, two languages

Debt and equity don't just price risk differently. They speak different dialects.

Bank and private credit conversations revolve around *what can go wrong*:

- collateral quality and liquidation value

- covenants that trigger early intervention
- cash flow coverage under stress
- legal rights, priority, and enforcement

Equity conversations revolve around *what can go right*:

- market size and the path to winning it
- unit economics that scale without breaking
- moats, not just momentum
- the team's ability to adapt when the plan changes (because it will)

CFOs get paid to be bilingual. You can be the most compelling storyteller in the world, but if your debt package is brittle, you'll be negotiating under pressure later. You can be the most conservative underwriter in the world, but if your equity story is lifeless, you'll raise too little or too expensively. The craft is matching the right capital to the right risk profile *and* telling the truth in the language the market understands.

A financing process is an auction even when you pretend it isn't

The biggest pricing lever in fundraising isn't your model. It's *competitive tension*. A single interested party is "a relationship." Two is "a process." Three is "a market."

This is not about gamesmanship. It's about protecting your operating timeline. When a lender knows they're the only option, your "fast close" becomes their "slow diligence," and your term sheet becomes a draft that mysteriously accumulates tighter covenants every week. When an investor knows you have alternatives, diligence still happens, but the calendar behaves.

Design the process like a CFO:

- a clear timeline (with real internal deadlines, not aspirational ones)
- standardized materials so every counterparty underwrites the same facts
- a defined decision-maker on your side (nothing kills momentum like "we'll get back to you after the next internal meeting")
- a small number of parallel conversations, not a sprawling beauty contest that drains management attention

A fundraising process that consumes the entire executive team for three months is not "thorough." It's expensive.

Credit agreements: where cheap money learns to say ``no"

A credit agreement is not a rate and a maturity. It's a rulebook for what happens when reality diverges from the plan. In good times, it sits quietly in a drawer. In bad times, it becomes the most important document in the building.

If you want to understand debt underwriting quickly, focus on three questions the lender is trying to answer:

1. *How do we get repaid?* (cash flow, asset sales, refinancing capacity)
2. *How do we see trouble early?* (reporting, covenants, KPIs)
3. *How do we control the downside?* (collateral, restrictions, remedies)

Your job is to make repayment boring, early-warning signals fair, and downside controls survivable.

Covenants: the difference between a partnership and a panic

Covenants are often described with the tone people reserve for dental surgery. That attitude is understandable and also misguided. The covenant package is the *price* of debt flexibility. A covenant that's too tight is effectively a hidden interest rate, because it forces renegotiation at the worst possible moment.

Covenants usually come in two families.

Maintenance covenants (always-on tripwires). These are tested periodically (often quarterly) and require you to stay within thresholds such as:

- maximum leverage
- minimum fixed-charge or interest coverage
- minimum liquidity

Maintenance covenants are valuable when they match the business cycle and include real headroom. They are dangerous when they are calibrated to a forecast rather than to a plausible downside.

Incurrence covenants (only tested when you take an action). These limit actions like additional debt, acquisitions, dividends, or asset sales unless you meet a test. Incurrence covenants give more day-to-day breathing room but constrain strategic moves.

A CFO's negotiating posture should be consistent: you're not trying to "avoid covenants." You're trying to avoid *accidental default*. The best covenant is one that only triggers when something meaningful is wrong—not when a customer payment slips two weeks.

Headroom is not a nice-to-have; it's an operating asset

Headroom is the gap between your expected performance and your covenant threshold. That gap has a cash value even though it doesn't show up on the balance sheet. It determines whether you can keep investing during a soft patch or whether every decision becomes lender choreography.

Here is a CFO-grade way to think about headroom:

- *Forecast headroom* is what you think you have.
- *Stress headroom* is what you actually have.

Stress headroom is built using the same discipline you use for liquidity planning: what happens if EBITDA drops, working capital swings against you, and the timing of receipts shifts? If your covenant compliance depends on "we will definitely hit the plan," you do not have headroom. You have a motivational poster.

The most dangerous sentence in lending: ``EBITDA is adjusted"

Most credit agreements define EBITDA in excruciating detail, because it's the denominator in leverage ratios and the numerator in coverage tests. Every add-back is a negotiation. Some are reasonable. Some are wishful. The risk is that you build a covenant package that relies on add-backs you can't defend when performance deteriorates.

A lender will tolerate adjusted EBITDA when adjustments are:

- truly non-recurring
- small relative to the base
- clearly documented

A lender will become skeptical when "non-recurring" appears every quarter like clockwork.

The CFO move here is both tactical and cultural: create an internal definition of "underwriting EBITDA" that is more conservative than "management EBITDA," and use it for covenant forecasting. You can still negotiate the credit agreement definition, but you won't be surprised by it.

Collateral and guarantees: the quiet terms that shape your future

Collateral isn't just about reducing the lender's loss given default. It's also about your future financing flexibility. Pledge everything now, and you may find that later lenders have nothing left to secure their loans, raising cost or limiting size. Guarantees can also pull subsidiaries into the credit net, affecting operational separability and M&A options.

The practical discipline:

- understand which assets are essential to operations and which are financeable without crippling flexibility
- model the *operational* impact of springing liens, blocked accounts, dominion of funds, and cash sweep mechanics

A low interest margin paired with aggressive control of cash can behave like a high interest margin, because it deprives you of discretionary liquidity exactly when you need it.

Equity fundraising: you are not selling shares; you are selling a believable future

Equity investors do not fund spreadsheets. They fund *the gap* between today and a future they can imagine—and they price the risk that you won't cross it.

The biggest mistake CFOs make in equity processes is presenting a plan that reads like a wish list: hire these people, build these features, enter these markets. That's activity. Investors want causality. They want to know which levers drive outcomes and why you have reason to believe those levers are real.

A compelling equity story has two layers:

- *The narrative:* what changes in the world make this company inevitable (or at least plausible)?

- *The math:* unit economics, retention, payback periods, and operating leverage that show the narrative can become cash flow.

If the narrative is strong but the unit economics are weak, you're selling charisma. If the unit economics are strong but the narrative is weak, you're selling a vending machine. The winning story is when the economics make the narrative feel *inevitable.*

The data room is your credibility factory

A data room is not a folder of documents. It's a test of how you run the company. Investors and lenders use it to infer your operational maturity:

- Do the numbers tie out across financial statements, KPI dashboards, and board materials?
- Are customer contracts organized, executed, and consistent?
- Can you explain churn, cohort behavior, margin moves, and working capital with clarity?
- Do you have clean cap tables, option grants, and board approvals?

Most diligence questions are not information requests; they are *confidence probes*. When you answer quickly, consistently, and transparently, you are doing more than moving a process along. You are reducing perceived risk, which is another way of saying you are lowering your cost of capital.

A practical tip that pays back immediately: build a "single source of truth" pack that includes your latest monthly financials, a rolling 13-week cash forecast, KPI definitions, and a bridge from GAAP to any non-GAAP metrics you use. When those documents exist and agree with each other, diligence becomes verification instead of investigation.

Term sheets: the headline price is the beginning, not the end

CFOs love clean comparisons. Term sheets resist them. Two offers can have the same valuation and radically different economics because of structure and rights.

On the debt side, watch the "rate" distractions. The true cost of debt includes:

- upfront fees and OID (original issue discount)
- amortization profile (cash burn disguised as "repayment")
- covenant tightness and the cure rights (or lack thereof)
- mandatory prepayments and cash sweeps
- reporting burden and audit requirements

The best deal is often not the cheapest on paper. It's the one that lets operations breathe while still being disciplined.

On the equity side, terms shape governance and outcomes. Venture and growth equity rounds often use a standardized document stack (with market norms heavily influenced by model documents), which makes negotiations feel familiar—and also makes it easy to miss landmines. Focus on the terms that change incentives and control:

- liquidation preferences (who gets paid first, and how much)
- participation (whether preferred also shares in the upside)
- anti-dilution mechanics
- protective provisions and board composition
- redemption rights and dividend features

A "great valuation" paired with harsh preferences can produce a future where the common shareholders celebrate revenue growth while the preference stack quietly eats the exit.

SAFEs and convertible instruments: speed now, math later

Early-stage fundraising instruments like SAFEs and convertible notes exist for a reason: priced equity rounds can be slow and expensive when the company is still finding its shape. These instruments trade negotiation speed for future complexity.

They are not harmless. They are deferred pricing decisions that can produce surprising dilution at conversion, especially when multiple rounds stack, caps vary, and "most favored nation" clauses propagate.

The CFO's discipline with convertibles is simple:

- maintain a conversion model that is updated after every new instrument
- show dilution under multiple valuation scenarios (including a "down round" case)
- communicate clearly with the CEO and board about what has been sold economically, not just legally

If you don't model conversion, you are fundraising with your eyes closed.

Regulatory reality: raising money is also compliance

In the U.S., the path you choose to raise capital determines what you can say, who can invest, what you must file, and what ongoing obligations you inherit.

Rule 506(c) (general solicitation under Regulation D). This pathway can allow broad solicitation, but it comes with a hard boundary: purchasers must be accredited investors, and the issuer must take reasonable steps to verify that status. Securities sold are restricted, and filings such as Form D have timing requirements. The tactical implication is that "marketing the raise" is not just a brand exercise; it's a compliance program you need to run cleanly.

Regulation Crowdfunding (Reg CF). Crowdfunding can open access to a broader set of investors, but it carries structured disclosure and ongoing reporting expectations. It's not "free marketing capital." It is a public-facing fundraising method with documentation, updates, and post-raise obligations that your finance team must be prepared to support.

Registered offerings and going public. Public markets can provide scale and liquidity, but they also impose a different operating cadence: registration statements, SEC effectiveness, and then ongoing reporting obligations. The real cost isn't just legal fees. It's the permanent shift to living in a disclosure regime.

The CFO takeaway is not "avoid complexity." It's "price complexity." Every compliance obligation is work that someone must do correctly, every quarter, in good times and bad.

Negotiation posture: the CFO's job is to keep the company fundable next time

A successful raise is not the one that closes. It's the one that leaves you:

- enough cash runway to hit a value-creating milestone, not just survive
- terms that don't choke operations at the first hiccup
- a cap table and governance structure that still attracts future capital
- credibility intact with stakeholders

This is where CFO temperament matters. You are the designated adult in a room full of optimism. Your job is to protect the company from financing terms that feel painless today and painful forever.

A good internal habit: treat every term sheet like a future case study. Imagine it's two years later, performance is 15% below plan, and you're explaining to the board why the terms you accepted are now constraining decisions. If the explanation sounds like "we didn't think that would happen," the term was probably too tight.

Actionable takeaway

Before you send a single deck to a lender or investor, build an "Underwriter Packet" that you can reuse across processes:

- last twelve months financials and trailing KPIs with clear definitions
- a rolling 13-week cash forecast (weekly granularity) and a 12-month cash bridge
- a conservative downside case showing liquidity, covenant compliance (if applicable), and management actions with timing
- a clean cap table with option pool details and board approvals
- a one-page narrative that links the use of proceeds to specific milestones and value drivers

If you can produce that packet in 48 hours, you are not just "prepared." You are negotiating from strength, because speed and clarity are credibility—and credibility is cheaper capital.

5.4 Keeping the Trust: Stakeholder Communication

Capital has a long memory. It remembers the quarter you missed your covenant by a hair and called the bank *after* the test date. It remembers the investor update that quietly stopped arriving. It remembers the board meeting where the forecast changed three times in forty-five minutes and nobody could explain why. Money doesn't only price your business model; it prices your *reliability*.

The paradox is that trust is most valuable when you least feel like earning it. When performance is strong, communication feels like celebration. When performance is weak, communication feels like exposure. CFOs who lean into disclosure in the messy moments end up with more flexibility, better terms, and a calmer organization. CFOs who delay tend to discover that silence is not neutral. Silence is interpreted as concealment, and concealment is priced as risk.

Trust is an operating system, not a personality trait

Some leaders are naturally charismatic. That helps. It's also not the main ingredient. Stakeholder trust is mostly operational: cadence, consistency, and a habit of saying the same truth in the same way over time.

The easy version of trust is "they like us." The durable version is "they believe our numbers." The durable version is built through routines:

- predictable reporting cycles
- consistent definitions for KPIs
- clean reconciliation between management metrics and GAAP (or your accounting standard)
- early escalation of risks, before they become events

The CFO owns these routines the way an engineer owns uptime. Nobody applauds a system that works. Everyone notices when it doesn't.

The *No Surprises* rule: the only rule that matters when things go wrong

There is a single principle that separates "normal volatility" from "relationship damage": no surprises. Not "no bad news." Bad news happens. The breach is when stakeholders learn it late, learn it sideways, or learn it with a story that keeps changing.

A surprise creates a second crisis on top of the first:

- the operational problem (missed number, lost customer, margin compression)
- the credibility problem ("what else don't we know?")

The credibility problem is usually more expensive. Operational problems can often be fixed with pricing, cost actions, or time. Credibility problems are financed at penalty rates.

The CFO's job is to make the first problem the only problem.

Three audiences, three anxieties

Boards, lenders, and investors ask different questions because they are exposed to different risks.

The board worries about stewardship.

They are responsible for oversight, strategy, and leadership continuity. Their fear is not just "are we missing the plan?" but "are we in control of the business, or is the business in control of us?" They want clarity on trade-offs: growth versus cash, risk versus resilience, speed versus certainty.

Lenders worry about downside and timing.

They care less about your five-year vision and more about whether you will violate a covenant in the next two quarters. Lenders want early warning, clear math, and practical remediation plans. They dislike narratives that sound like hope, especially when the cash forecast says otherwise.

Equity investors worry about upside and integrity.

They can tolerate volatility if they believe the long-term story is intact and management is candid. The instant they suspect selective disclosure or moving goalposts, they start treating every claim as marketing. A company can survive a missed quarter; it struggles to survive being labeled "unreliable."

If you speak to all three audiences with the same script, you will confuse at least two of them. The CFO's craft is to keep the facts consistent while tailoring the emphasis to the specific anxiety in the room.

Cadence beats brilliance: the power of regular disclosure

A surprising number of communication disasters begin with a well-intentioned plan to "only update people when there's something meaningful." That feels efficient. It also trains stakeholders to assume that *every* communication is crisis-coded. When you finally do reach out, they brace for impact.

Regular disclosure has an underappreciated advantage: it makes your communications emotionally boring. Boring is good. Boring means trust is being maintained without drama.

A practical cadence that works in many companies:

- *Board:* monthly financial package (even if meetings are quarterly), with consistent dashboards and a short memo highlighting variances and decisions required.
- *Lenders:* covenant compliance package on the schedule required, plus proactive heads-up calls when headroom shrinks materially.
- *Investors:* a monthly update for private companies, quarterly earnings and event-driven disclosures for public ones.

The key is that cadence is not "reporting." It's relationship maintenance. When the cadence is steady, the occasional emergency message doesn't feel like an ambush.

What to disclose: the art of being candid without being chaotic

"Transparency" is often misunderstood as "tell everyone everything." That's not feasible, and it can be irresponsible. The goal is not maximal disclosure. The goal is *decision-useful* disclosure.

Decision-useful disclosure has three traits:

- *Comparable:* same metrics, same definitions, over time.
- *Complete:* the number and the driver, not just the number.
- *Consequence-aware:* what it means for liquidity, covenants, runway, hiring pace, or strategic options.

This is where CFO discipline shows up. When stakeholders sense that management is throwing data at them to avoid clarity, they stop trusting the data. When they sense that management is simplifying so aggressively that nuance is lost, they stop trusting the story.

A good rule: if a stakeholder asks the same question twice in two months, it's not a stakeholder problem. It's a packaging problem.

Bad news should arrive early, with math, and with choices

The best way to deliver bad news is to make it feel less like tragedy and more like governance. That requires three ingredients.

Early.

Early means before a covenant test date, before cash gets tight, before a quarter closes. "Early" is when you still have options. After that, you have negotiations.

With math.

Not performative math. The real drivers: revenue, margin, working capital, capex, and timing. Stakeholders can tolerate pain; they struggle with mystery.

With choices.

The worst update ends with "we'll work harder." The best update ends with "here are the levers, here's the cost of pulling them, and here's what we recommend." Choices restore a sense of control.

A CFO who can frame a difficult quarter as a set of controlled trade-offs will keep stakeholders engaged. A CFO who frames it as an emotional event will drain confidence.

Lender communication: treat them like risk managers, not adversaries

Lender relationships go bad in a predictable way. First, the company starts managing the lender *as a compliance exercise*. Then performance softens, and the company becomes reluctant to share partial information. Then the lender discovers the problem late, feels blindsided, and tightens terms or freezes flexibility. The company interprets that as hostility. Everyone's time gets consumed.

The way out is to treat the lender as what they are: a professional risk manager.

A strong lender update typically includes:

- a clear view of trailing performance and variance to budget
- forward-looking liquidity, with explicit assumptions
- covenant headroom projections under base and downside cases
- management actions already taken (not just planned)
- specific asks, if any (waiver, amendment, additional capacity), with rationale

Two tactical moves matter more than they should:

- *Call before you email.* A short conversation lowers the temperature and reduces misinterpretation.

- *Don't sandbag the model.* If your forecast is unrealistically pessimistic to "create cushion," lenders will assume you're hiding upside or that the business is unstable. Use a credible base case and a credible downside case instead.

The lender does not need optimism. They need predictability.

Board communication: make oversight easy, not theatrical

Board members are smart, busy, and often managing multiple portfolios of risk at once. The CFO's job is to make the company legible.

That means:

- a consistent board package that ties financials, KPIs, and cash together
- a short written narrative that explains what changed and why
- explicit decisions required from the board (and what happens if no decision is made)

A board meeting should not feel like a forensic investigation. If the board is spending its time reconciling numbers, it cannot spend its time on strategy.

A subtle but powerful move: show a bridge from last month's forecast to this month's forecast. Not a new plan dropped from the sky—a bridge. Forecast bridges make management accountable without becoming punitive. They also expose whether the organization is learning or merely reacting.

Investor communication: the line between storytelling and selective disclosure

Equity investors fund possibility, but they punish inconsistency. The most effective investor updates are not the ones full of good news; they're the ones that feel *true*.

Truth has structure:

- what you said would happen
- what happened
- why the difference occurred
- what you're doing about it
- what you now believe will happen next (and what could change that belief)

For private companies, investor updates are often informal. That informality is dangerous if it becomes sloppy. A casual tone is fine. Casual facts are not.

For public companies, the line is sharper because the rules are sharper. Regulation FD (Fair Disclosure) in the U.S. is built on a simple idea: if material nonpublic information is intentionally disclosed to certain people (like analysts or institutional investors), it must be disclosed publicly as well; if it's disclosed unintentionally, public disclosure must follow promptly. The practical implication is not just "be careful." It's "build a communications process that prevents accidental selectivity."

One of the easiest ways to violate trust (and sometimes rules) is to "clarify" performance privately. If a clarification changes how an investor would value the company, it probably belongs in a broad disclosure channel, not a quiet call.

Event-driven disclosure: the 8-K mindset even when you're private

Public companies live with a discipline that private companies can borrow: event-driven updates. In U.S. public markets, Form 8-K is the mechanism for disclosing certain material events on a timely basis, and it sits alongside periodic reports like the 10-Q and 10-K. The point isn't the form itself. The point is the muscle: when something material happens, you communicate it promptly, consistently, and with context.

Private companies often do the opposite: they wait until the next scheduled board meeting or investor update to "roll it in." That's convenient for management and often corrosive to trust.

An "8-K mindset" for any company looks like:

- define what counts as a material event for your stakeholders (customer loss, covenant headroom collapse, CEO transition, major litigation, significant financing changes)
- pre-draft disclosure templates so the first draft isn't written under adrenaline
- decide who approves the message and what channels it goes through

When you pre-wire the response, you reduce the chance that fear drives silence.

Communication channels: your process must match your fundraising method

Even outside public markets, channels impose constraints.

If you raise under structures that allow broad solicitation, you may still have conditions linked to who can invest and how eligibility is verified. If you raise via crowdfunding regimes, communications can carry specific requirements and ongoing reporting obligations. These frameworks have a common theme: you cannot treat stakeholder communication as improvisation when the rules expect repeatable disclosure.

The CFO's practical move is to treat communications as part of your control environment:

- centralize investor communications drafts
- keep an archive of what was disclosed and when
- align external messaging with internal reporting so you never have two versions of reality

This isn't bureaucracy. It's future-proofing. When the company grows, the standards rise. The habits you build early determine whether that transition is smooth or traumatic.

The most credible sentence in finance: ``Here's what we don't know yet"

Many leaders think confidence means certainty. Stakeholders think confidence means *self-awareness*. A CFO who can clearly separate facts from hypotheses is far more believable than one who speaks in absolutes.

The trick is to pair uncertainty with a plan:

- what is uncertain
- what information will resolve it
- when that information will be available
- what you will do in the meantime

This converts ambiguity from a threat into a managed process. It also inoculates you against "moving the goalposts," because you were explicit about what could change.

When you miss: the recovery playbook is communication plus action

Trust can survive a miss. It rarely survives a miss followed by obfuscation.

When you miss guidance, breach an internal target, or face a covenant risk, the recovery sequence matters:

- communicate quickly, with a clear diagnosis
- show immediate containment actions (cash preservation, spend controls, operational focus)

- present a realistic re-forecast with assumptions stakeholders can interrogate
- define leading indicators that will prove whether the fix is working before the next quarter ends

Leading indicators are a gift to stakeholders because they reduce the time they spend living in uncertainty. If the only proof you offer is next quarter's results, you're asking them to wait in the dark.

Actionable takeaway

Create a "No Surprises Protocol" that is short enough to be used under stress:

1. Define five events that require immediate stakeholder notification (tailored separately for board, lenders, and investors).
2. For each event, pre-write a one-page template: facts, impact on liquidity/covenants/runway, actions taken, actions planned, and the next update date.
3. Set a standing internal rule: if covenant headroom or cash runway deteriorates by a meaningful amount versus the prior forecast, the CFO schedules the call *within 48 hours*.

Run the protocol once as a drill. If it feels awkward, that's the point. You're practicing the moment when trust is either preserved by professionalism or lost to delay.

Chapter 6

Strategic Allocation: Valuation and M&A

Making money is hard work, but spending it wisely is an art form. Once a company generates cash, the CFO becomes the Chief Allocation Officer, tasked with a high-stakes decision: do we reinvest in our own machinery, buy a competitor, or just save for a rainy day? This chapter moves beyond simple budgeting to the strategic bets that define a company's future. We will explore the rigorous math of internal investments, the subjective science of valuing businesses, and the rollercoaster reality of Mergers and Acquisitions-where the real work begins only after the deal is signed.

6.1 Investing Internally: The Discipline of CapEx

A capital expenditure request is rarely presented as what it is: a bet. It arrives dressed up as a "need" ("the plant is at capacity"), a "risk reduction" ("the current system is brittle"), or a "strategic must" ("our competitors have already automated"). But under the suit, it's still a bet on future cash flows—and a bet you can't easily unwind. You can't return a warehouse to the store. You can't "cancel" an ERP implementation halfway through and expect your team to emerge emotionally unscathed. That irreversibility is why CapEx deserves governance that's more disciplined than the average operating expense, and more skeptical than the average strategic slide deck.

Most companies don't get hurt because they don't know how to calculate Net Present Value. They get hurt because they let projects *skip the part where reality is invited into the room*. The problem isn't the math; it's the psychology. Sponsors fall in love with an idea, managers feel pressure to show growth, and teams quietly round assumptions in the direction that makes the spreadsheet look brave. The CFO's job isn't to be the person who says "no." It's to be the person who makes "yes" mean something.

CapEx governance is a speed limit, not a brake

The strongest governance frameworks don't slow the business down; they stop it from crashing at high speed. Think of CapEx governance as a set of speed limits and guardrails that let you move fast *without* letting enthusiasm turn into negligence. Good governance does three things simultaneously:

- forces clarity about what success looks like (in cash, not adjectives),
- creates comparability across very different projects (a new line vs. a new CRM),
- and builds an institutional memory so the company gets smarter over time.

The practical trick is to treat governance like an operating system, not a one-off approval meeting. The meeting is the visible part. The operating system is everything that happens before and after: how ideas are shaped, what templates force hard choices, which metrics are mandatory, how risk is expressed, and how the organization learns when the outcome doesn't match the pitch.

A clean governance model begins with a simple principle: *the sponsor owns the story; finance owns the skepticism; the executive committee owns the trade-off*. If the sponsor is also the scorekeeper, you will eventually approve a beautiful tragedy. If finance is also the decision-maker, you'll get accused of being anti-growth and the business will route around you. The healthiest model separates roles and makes the trade-offs explicit.

The Greenlight Process: turning a passionate idea into a financeable bet

A "Greenlight Process" sounds bureaucratic until you remember the alternative: the organization funds projects through charisma, urgency, and whoever cornered the CEO first. The point of a structured process is not to standardize creativity. It's to standardize *evidence*.

A practical Greenlight Process has stages, and each stage has a purpose:

- *Idea shaping*: convert a vague desire into a defined scope and measurable outcomes.
- *Pre-screen*: determine whether this is even worth modeling (and whether it belongs in CapEx).
- *Business case*: produce a decision-ready model with explicit assumptions, risks, and alternatives.
- *Approval*: allocate scarce capital across competing bets.
- *Post-audit*: compare promised vs. delivered and capture lessons while they still sting.

If this feels heavy, remember that the organization is already paying for a process. It's just paying in a different currency: rework, sunk-cost paralysis, and the quiet tax of systems that never quite deliver what was promised.

The CFO can make the process feel like an advantage rather than a hurdle by offering something every sponsor wants: a fair fight. A consistent template means the project is evaluated on its merits, not the politics of the presenter. A clear hurdle rate means the sponsor isn't guessing what the committee "likes." And a post-audit means the organization stops repeating the same optimistic mistakes with fresh confidence each time.

NPV: the truth-teller that speaks in dollars

Net Present Value (NPV) is the most useful tool in capital allocation because it answers the only question that matters: *after paying for this project and for the risk of waiting, how many dollars did we create (or destroy)?*

Here's the plain-language definition: NPV converts future cash inflows and outflows into today's dollars using a discount rate (your required return), then subtracts the upfront investment. If the result is positive, the project is expected to create value *in excess* of the return your capital could earn elsewhere at comparable risk.

NPV is unpopular for one reason: it refuses to be flattering. It doesn't care that a project is exciting, visible, or politically sponsored. It cares about timing and magnitude of cash. And it punishes delays like a metronome. A benefit that arrives in year five is worth less than the same benefit arriving in year two, even if the PowerPoint makes them look equally shiny.

The CFO's leverage comes from making NPV feel less like a finance ritual and more like a shared language. A sponsor doesn't need to become a valuation expert. They need to understand three intuitive levers:

- *Size*: bigger net cash benefits increase NPV.
- *Speed*: earlier benefits increase NPV disproportionately.
- *Certainty*: riskier cash flows should be discounted more heavily, reducing NPV.

If you want to improve project quality quickly, stop arguing about whether the model is "right" and start asking which lever the sponsor is relying on. "Is this project valuable because it's big, because it's fast, or because it's unusually certain?" The answer tells you where to push.

IRR: a useful metric that can quietly lie

Internal Rate of Return (IRR) is often treated like the gold medal of capital budgeting because it comes out as a percentage. People love percentages. A 28% IRR sounds heroic, even if it only applies to a small project that moves no needles. The plain-language definition: IRR is the discount rate that makes the project's NPV equal zero. In other words, it's the "break-even return" implied by the cash flows.

IRR is not useless; it's just frequently misused. It's helpful when you're comparing projects of similar scale and similar timing, or when capital is constrained and you need a sense of efficiency. But IRR can mislead in three common ways:

- *Scale bias*: a tiny project can have a spectacular IRR and still create trivial dollar value; a large project can have a lower IRR and create far more value.
- *Timing games*: pushing costs later (or benefits earlier) can inflate IRR even if the underlying economics aren't better.
- *Reinvestment fantasy*: IRR implicitly assumes interim cash flows can be reinvested at the same IRR, which is often unrealistic for very high IRRs.

The CFO move is to keep IRR on a short leash. Use it as a secondary signal, never as the deciding vote. In committee, you want the conversation to sound like: "This project has a 19% IRR, which clears the hurdle, and it creates $12M of NPV. The alternative creates $9M of NPV. If we can only fund one, we should fund the $12M." That's adult decision-making: dollars first, percentages second.

The discount rate: where strategy and finance shake hands

The discount rate looks like an abstract finance input until you realize what it represents: your organization's opportunity cost for taking risk and waiting. It's the price of capital, in time.

A sloppy discount rate policy turns governance into theater. If every project uses an arbitrary 10% "because that's what we always use," you'll systematically over-invest in risky projects (because you didn't penalize risk) and under-invest in safe, steady cash savers (because you overpriced certainty). A strong policy does two things:

- establishes a baseline corporate hurdle rate (often tied to the company's weighted average cost of capital, adjusted for practical decision-making),
- and applies risk adjustments for projects that are meaningfully more uncertain than the core business.

Risk adjustment doesn't have to be a PhD exercise. You can implement a pragmatic tiering system: "core replacement," "capacity expansion," "new product adjacent," "new market / new tech." Each tier has a standard premium unless the sponsor can argue for an exception. The key is consistency. The moment discount rates become negotiated project-by-project, the process turns political and the most persuasive presenter wins.

One subtle governance win: make the discount rate policy explicit in the business case template, and require that any deviation is approved by finance *before* the model is built. This prevents the awkward situation where a sponsor builds a gorgeous model around a too-low hurdle and then you become the villain for changing it at the end.

Incremental cash flows: the difference between decision-useful and spreadsheet art

Capital budgeting models fail most often not because the formula is wrong, but because the cash flows are not *incremental*. Incremental means: the cash flows that change *because you do the project*, compared to the best realistic alternative.

This sounds obvious until you sit in the meeting where someone includes "allocated corporate overhead" as a project cost (which makes good projects look bad), or includes revenue that would have happened anyway (which makes bad projects look good). If you want a single sentence to tattoo on the business case template, it's this: *Model the delta between the world with the project and the world without it.*

The "without" world is not the current world frozen in amber. It's the most likely path the business will take if you *don't* fund this project. That's why alternatives matter. A new distribution center should be compared not just to "do nothing," but to leasing third-party capacity, expanding shifts, or redesigning the network. The discipline of alternatives is what prevents a sponsor from presenting a single option as inevitable.

The same logic applies to "sunk costs"—money already spent. If you've already spent $2M on preliminary engineering, that money is gone whether you proceed or not. Including it in the NPV calculation doesn't make the project more honest; it makes the decision worse by encouraging sunk-cost fallacy. You still document sunk costs, because they matter for learning and accountability. You just don't let them vote.

Payback: the comfort metric that can sabotage value

Executives often ask for payback period because it feels like common sense: "How fast do we get our money back?" In a world where forecasts are uncertain and careers are finite, payback feels emotionally safe.

The problem is that payback quietly teaches the organization to prefer projects that return cash quickly even if they create less value overall. It also ignores everything that happens after payback. A project that pays back in two years and then delivers mediocre cash for a decade can look better on payback than a project that pays back in three years and then delivers extraordinary cash for a decade.

You don't have to ban payback. You just have to demote it. Treat payback as a liquidity and risk lens, not a value lens. For example:

- use payback to flag projects that keep cash tied up too long,
- use NPV to decide whether the project is worth doing,
- and use scenario analysis to understand what could break the economics.

This framing keeps leaders comfortable without letting comfort overrule value.

Sensitivity analysis: forcing the model to admit what it depends on

A business case without sensitivity analysis is a story told in one voice. Sensitivity analysis forces the story to meet its critics.

In practice, you don't need a rainbow tornado chart for every project. You need a disciplined way to answer: *what assumptions are doing the heavy lifting?* If the NPV goes from +$8M to -$3M when utilization drops by five points, then utilization is not a detail; it's the project's heartbeat. That's where governance should focus: not on arguing about decimal places, but on validating the few variables that can flip the decision.

A useful rule: every major project should show at least three scenarios:

- *Base case*: sponsor's most likely view, with assumptions documented.
- *Downside case*: the world where the main risks actually happen (delays, lower adoption, higher costs).
- *Upside case*: the world where execution is excellent and tailwinds show up.

Then the committee should ask a more interesting question than "Is the base case positive?" Ask: "In the downside case, do we survive without regret?" If the downside case implies operational chaos or covenant stress, the project may still be worth doing—but it's no longer a simple "approve." It becomes a risk-managed program with phased funding, gates, and contingency plans.

Stage gates: buying information before buying the whole project

One of the most practical governance tools is staged approval. Instead of approving $30M upfront for a multi-year project, approve $2M for design and vendor selection, then $8M for pilot and site preparation, then the remaining $20M when key uncertainties have been reduced.

This does two powerful things. First, it aligns funding with learning. Second, it gives the organization permission to stop. Many companies claim they want to stop bad projects early. Their approval mechanics say otherwise: once the full budget is approved, stopping becomes socially equivalent to admitting failure. Stage gates make stopping feel like *successfully completing a phase*, not publicly quitting.

Stage gating also improves negotiation leverage with vendors. When the business case includes a clear decision point after pilot results, vendors know they have to perform to win the next tranche. Governance becomes a tool for procurement and execution, not just finance.

The committee: how to keep it from becoming a courtroom

Capital committees often drift into one of two bad personalities: the rubber stamp or the courtroom. The rubber stamp approves everything because no one wants to be the person who slows growth. The courtroom turns every meeting into adversarial theater, so sponsors learn to lawyer their cases rather than improve them.

The CFO can shape the committee into something better: a place where capital is treated like inventory. Scarce, valuable, and allocated with intent.

A few design choices matter more than any policy document:

- *A single scorecard*: every project is presented with the same small set of metrics (NPV, IRR, payback, strategic rationale, key risks, capacity/people constraints).
- *Comparative framing*: projects are discussed in relation to alternatives *and* to other uses of capital, not in isolation.
- *Time-boxed debate on assumptions*: committees should argue about the two or three drivers that matter, not the formatting of the model.

The most effective committees also create an explicit category for "mandatory" spend (safety, compliance, true asset replacement) and evaluate those differently from growth bets. Mixing them without acknowledging the difference creates a false debate: no one wants to argue against compliance, so growth projects hide under compliance language. Separate buckets restore honesty.

The post-mortem audit: where the organization earns compound interest on experience

The post-mortem audit is the most underused lever in CapEx governance, and it's the closest thing finance has to a superpower. Without it, every project is a one-time argument. With it, every project becomes training data.

A post-mortem audit isn't a witch hunt. It's a reconciliation between promises and outcomes. The right tone is forensic, not punitive: "What did we think would happen? What actually happened? Why? What should we do differently next time?"

To make this work, the audit must be designed into the process at the beginning, not bolted on at the end. The business case should include:

- the few metrics that define success (not twenty),

- a measurement plan (where the data will come from),
- and a date when the project will be evaluated (often 6, 12, and 24 months after go-live).

Then finance has to do the unglamorous work: build a mechanism to track actual results against the model *separately from the noise of the business*. If a project promised $4M in annual labor savings, you need a way to see whether labor cost actually dropped, whether headcount changed, whether overtime fell, or whether the savings were absorbed by growth elsewhere. Otherwise the organization declares victory by vibes.

Post-mortems also reveal a deeper truth: many "failed" projects are really forecasting failures, scope creep failures, or change-management failures. That distinction matters. If the economics were sound but adoption lagged, the fix is not "stop investing." The fix is strengthening execution muscles: training, incentives, process redesign, and accountability.

One more uncomfortable benefit: post-mortems deter strategic misrepresentation. When sponsors know their assumptions will be revisited in a year, the temptation to smuggle optimism into the model declines. Governance becomes self-enforcing.

A CFO-grade business case template (what it forces, not what it contains)

A good template doesn't just capture information; it forces decisions. It should be short enough that people actually use it and strict enough that it prevents the most common games.

At minimum, a CFO-grade template forces five acts of honesty:

1. *What problem are we solving, in measurable terms?* (Not "modernization"; something like "reduce order cycle time by 24 hours" or "avoid 8% annual downtime growth.")
2. *What is the alternative?* (The best realistic "no" path.)

3. *What are the incremental cash flows, and where do they come from?* (Savings are not benefits unless they translate into cash or avoid real spend.)
4. *What must go right?* (A short list of key assumptions and risks, with owners.)
5. *How will we measure success after go-live?* (Defined metrics, data sources, and timing.)

Notice what's missing: the lengthy narrative trying to prove the sponsor is a good person with good intentions. Intentions don't pay back capital. Cash does.

Making CapEx decisions in a world where people are tired

A final reality check: capital allocation happens inside organizations full of humans with full calendars. The smartest governance framework fails if it requires heroic effort to comply.

So make it easier to do the right thing than the wrong thing. Prebuild model shells that already handle depreciation, taxes, working capital, and discounting correctly. Provide a "drivers" tab where the sponsor only inputs a handful of operational assumptions. Standardize how benefits are translated into cash impact (for example, the difference between "labor productivity" and "labor cost reduction"). Create a finance partner model where corporate finance helps shape the case early rather than policing it late.

When finance shows up at the beginning, sponsors experience governance as assistance. When finance shows up at the end, sponsors experience governance as prosecution. Same math, completely different culture.

Actionable takeaway: install one rule that changes behavior tomorrow

Pick one change you can implement immediately: *no major CapEx approval without a dated post-mortem plan and three named success metrics tied to cash*. Not a paragraph promising accountability. A real plan: metrics, data sources, and the calendar invites for the 6/12/24-month reviews.

That one rule does something magical. It doesn't just improve measurement. It changes the *pitch*. Sponsors start crafting projects that can survive future scrutiny. Assumptions get cleaner. Scope gets sharper. And capital—which is always scarcer than optimism—starts flowing toward projects that can prove they deserve it.

6.2 What's It Worth? The Science of Valuation

Walk into an open house and you'll see two tribes immediately. One is wandering room to room, picturing future life: morning light in the kitchen, kids on the stairs, the roof that won't leak for a decade. The other is doing math on their phone: recent sales per square foot, property taxes, mortgage rates. Both tribes are trying to answer the same question—*what's it worth?*—and both are right to be suspicious of the asking price.

Businesses get valued the same way. One lens is personal and structural: what cash will this thing throw off over time, and how risky is that cash? That's intrinsic value, usually expressed as a discounted cash flow (DCF). The other lens is social and comparative: what are similar businesses trading for, and what have buyers recently paid? That's relative value, usually expressed as multiples (trading comps and precedent transactions). Valuation looks like black magic only when you pretend it's one number. It's not. It's a set of arguments, with math as the referee.

A CFO's advantage is that you don't have to *win* valuation. You have to *use* it. Valuation is the language that turns strategy into capital decisions: how much can we pay, what must be true for this to work, and what would make us regret it later. The goal isn't a perfect number; it's a decision you can defend in daylight.

Price vs. value: the most expensive confusion in corporate finance

Price is what the market (or a counterparty) demands today. Value is what the asset is worth to an owner, based on cash it can generate in the future. In calm markets those numbers flirt. In stressed markets they stop talking.

This sounds philosophical until you're the buyer. If you treat market price as value, you'll overpay in hot cycles and freeze in cold ones. If you treat your model's value as price, you'll lose good deals because reality refuses to match your spreadsheet. The competent stance is more practical: *value is your anchor; price is the weather*. You don't ignore the weather. You just don't let it convince you the ocean moved.

That framing also keeps internal conversations sane. Business leaders naturally fall in love with narratives. Investors naturally fall in love with comparables. Finance has to make both camps slightly uncomfortable—because discomfort is where the assumptions finally show themselves.

The two main families of valuation (and why you need both)

Intrinsic valuation asks: *what is this business worth if we own it and live with its cash flows?* DCF is the workhorse here. You forecast free cash flow (cash generated after paying operating costs and necessary reinvestment), then discount those cash flows back to today using a required return that reflects risk. It's the open-house tribe imagining the future life—except you're imagining cash.

Relative valuation asks: *what does the market pay for similar cash-flow stories?* Multiples take observed prices and scale them by a financial metric: earnings, EBITDA, revenue, sometimes gross profit. It's the comp-sales tribe—except the "three-bed two-bath down the street" is now "a SaaS company with 120% net revenue retention" or "a manufacturer with 18% EBITDA margins."

When these two methods agree, you get confidence. When they disagree, you get insight. Disagreement is not a failure; it's the whole point. It tells you whether the market is pricing something you're not modeling (momentum, scarcity, optionality) or whether your model is relying on assumptions the market doesn't believe.

DCF in plain language: paying today for tomorrow's cash

A DCF has a reputation for complexity because it has many lines, but the underlying idea is almost stubbornly simple: *value equals the present value of future cash.* Present value just means future dollars are worth less than today's dollars because you can invest today's dollars and because the future is uncertain.

The CFO-grade way to think about DCF is to break it into four decisions that are more judgment than math:

1. *What cash flows are we valuing?* (Free cash flow, not accounting earnings.)
2. *How long does abnormal performance last?* (The explicit forecast period.)
3. *What does "mature" look like?* (Terminal value assumptions.)
4. *How risky is the cash?* (Discount rate.)

If a model feels intimidating, it's usually because these decisions are hidden behind formulas. Pull them into the open and the model becomes a conversation rather than a rite.

Free cash flow: the cash that's actually available to owners

Earnings can be flattering; cash is harder to sweet-talk. Free cash flow (FCF) is the cash the business generates that could, in principle, be distributed to capital providers (debt and equity) after funding operations and reinvestment.

A common definition for free cash flow to the firm is:

$$\text{FCF} = \text{EBIT}(1 - T) + \text{D\&A} - \text{CapEx} - \Delta\text{NWC}$$

Each term carries a story:

- *EBIT*(*1-T*): operating profit after tax, ignoring financing choices.
- *D&A added back*: depreciation is real economics over time, but it's not cash this period.
- *CapEx*: the cash you must spend to keep the machine running and to grow it.
- Δ*NWC*: growth often consumes cash through inventory and receivables before it produces it.

A useful instinct: when someone tells you a business is "high margin," ask what happens to cash when it grows. Many businesses look gorgeous on EBITDA and quietly devour cash through working capital and maintenance CapEx. DCF forces that truth onto the page.

Discount rate: the price of risk and time

Discounting is where a DCF stops being arithmetic and starts being philosophy with a spreadsheet. The discount rate reflects the return investors demand for bearing risk. For valuing the entire enterprise (debt and equity together), the typical discount rate is the weighted average cost of capital (WACC).

You don't need to worship WACC. You need to respect what it's trying to do: match cash-flow risk with a required return. A stable regulated utility deserves a different discount rate than an early-stage consumer brand that lives and dies by next quarter's customer acquisition costs.

Here's the uncomfortable part: you can make almost any DCF "work" by nudging the discount rate. That's why governance matters. A credible process fixes the discount-rate methodology up front, then argues about operating drivers rather than quietly bargaining the discount rate to get the answer someone wants.

Terminal value: where most of the value hides

Terminal value is the value of the business beyond the explicit forecast period. It often represents the majority of a DCF's total value, which is both normal and dangerous: normal because businesses are supposed to be going concerns, dangerous because the terminal value is where hand-waving likes to set up camp.

Most terminal values come from one of two approaches:

- *Perpetuity growth*: assume cash flows grow at a modest constant rate forever.
- *Exit multiple*: assume the business sells at a market multiple in the terminal year.

Both are acceptable; both can be abused. The CFO move is to ask what "maturity" really implies for margins, reinvestment, and growth. A perpetuity growth rate above long-term nominal GDP is a quiet claim that the company eventually becomes the economy. That's not ambition; it's arithmetic vandalism.

Multiples in plain language: borrowing the market's judgment

Multiples look simpler than DCF because they compress complexity into one ratio. That's their power and their hazard.

A multiple is a pricing shorthand:

$$\text{Value} = \text{Metric} \times \text{Multiple}$$

The metric might be EBITDA, earnings, revenue, or free cash flow. The multiple comes from peers (trading comps) or past deals (precedent transactions). Multiples tell you what the market is paying for a certain *type* of cash-flow story right now.

Trading comps vs. precedent transactions

Trading comparables use public market prices for similar companies. They reflect minority positions, daily liquidity, and the market's mood. Precedent transactions use acquisition prices for similar companies. They reflect control, synergies (real or imagined), and negotiating dynamics—which often means they include a premium.

Both are useful. Trading comps answer, "What does the market think a standalone business like this is worth today?" Precedents answer, "What have buyers been willing to pay to own the whole thing?" If your acquisition price is below the middle of credible precedents, you might be getting a deal—or you might be buying something with problems the comps don't have.

Choosing a multiple is choosing a story

Multiples are not interchangeable. EV/EBITDA is popular because it's widely available and less sensitive to capital structure, but it can reward businesses that underinvest in CapEx (maintenance is a cash cost; EBITDA ignores it). EV/Revenue can be useful for high-growth businesses with negative earnings, but it smuggles in an assumption that revenue will become profit someday. P/E is intuitive but heavily affected by leverage and accounting.

A practical CFO habit is to use at least two metrics: one that ties to operating profitability (like EBITDA or EBIT) and one that anchors on scale (like revenue) or cash (like unlevered FCF), depending on the business model. If those methods imply wildly different values, you've learned something. Usually, you've learned where profitability is fragile or where growth requires heavy reinvestment.

Multiple	*Best used when (and the trap)*
EV/EBITDA	Mature or scaling businesses with meaningful operating profit. Trap: ignores maintenance CapEx and working capital; can flatter asset-heavy models.
EV/EBIT	Useful when depreciation reflects real economic cost (manufacturing, logistics). Trap: accounting policies can distort EBIT comparability.
EV/Revenue	High-growth or early-stage businesses where margins are still forming. Trap: pays for growth without proving unit economics.
P/E	Stable businesses where leverage is part of the model and earnings quality is high. Trap: capital structure and one-time items can make P/E misleading.
EV/FCF	Cash-focused view when reinvestment needs are visible and stable. Trap: short-term FCF can be boosted by starving CapEx or pulling working capital forward.

The bridge you must not skip: enterprise value vs. equity value

Valuation meetings often derail because two people are answering different questions without realizing it.

Enterprise value (EV) is the value of the operating business available to all capital providers. Equity value is what remains for shareholders after paying off net debt and other claims. If you value a business using EV multiples (like EV/EBITDA), you are implicitly estimating enterprise value. To get to equity value, you adjust for net debt (and often other items like preferred stock, minority interests, unfunded pensions, or non-operating assets).

This matters because deal prices are paid for equity (the shares) but often discussed in enterprise terms. A buyer might say, "We paid 10x EBITDA," while the CEO hears, "The business is worth 10x EBITDA." The hidden variable is the balance sheet. High leverage can turn a reasonable enterprise valuation into an equity price that is either painful (for buyers) or surprisingly low (for sellers).

A CFO who insists on an explicit EV-to-equity bridge in every valuation discussion prevents expensive misunderstandings and forces the balance sheet back into the conversation where it belongs.

Immediately after the methods: triangulation on one page

A competent valuation rarely produces a single number. It produces a range from each method, then a narrower range where the methods overlap and your assumptions feel defensible. Investment bankers popularized the "football field" chart for a reason: it stops people from obsessing over precision and starts the right debate about reasonableness.

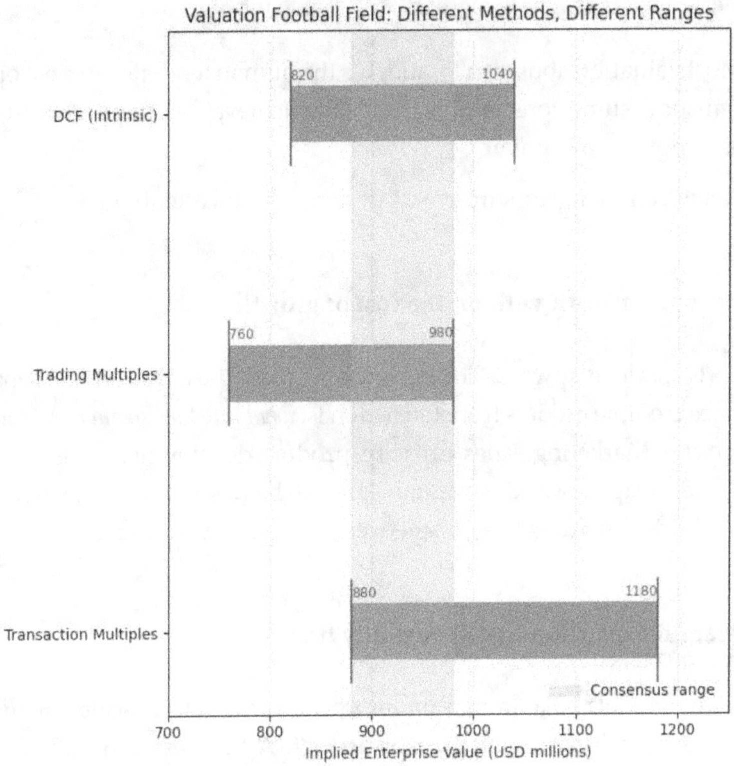

When someone insists "the valuation is $950M," point to the chart and ask a better question: *what would have to be true for $950M to be fair?* That question turns valuation from a tug-of-war into a due diligence roadmap. If $950M sits near the top of the DCF range, maybe the model assumes faster ramp, higher terminal margins, or lower reinvestment needs. If $950M sits near the bottom of transaction precedents, maybe the seller expects a control premium you're unwilling to pay. The number is less interesting than the assumptions that defend it.

Where valuation gets manipulated (often without malice)

Most valuation abuse isn't fraud. It's the human tendency to treat optimistic assumptions as "aggressive but achievable" and pessimistic assumptions as "unfair."

A few common pressure points deserve special attention:

Revenue growth without the cost of growth

Forecasts love upward-sloping revenue lines. The question that separates credible models from fan fiction is: *what must we spend to get that growth?* Marketing, sales capacity, product development, customer success, capex, working capital—growth has a shadow. If the model pretends growth is free, it will overvalue almost anything.

Margin expansion as a personality trait

Many models bake in margin improvement because "scale." Sometimes scale helps. Sometimes it doesn't. A business can scale and still face rising fulfillment costs, wage pressure, customer acquisition inflation, or price competition. Margin expansion should be tied to operational mechanisms: automation, mix shift, procurement leverage, pricing power, network effects. If the mechanism isn't described, the model is hoping.

Terminal value optimism

Terminal assumptions are where narratives go to hide. A small change in terminal growth or terminal multiple can swing value dramatically. A good discipline is to run the model with a deliberately sober terminal view, then earn your way back up with evidence: contract duration, switching costs, capital intensity, long-term market structure.

Synergies: the most seductive word in M&A

Synergies can be real. They can also be the sugar that helps an overpayment go down. The clean way to handle synergies is to model them explicitly (timing, costs to achieve, probability) and then decide how much of them you're willing to pay away to the seller. Paying for *your own* execution is a classic error. If you're the one who has to do the hard integration work, the value of that work should mostly accrue to you.

Even outside acquisitions, the synergy mindset shows up in internal valuation: "If we build this platform, everything gets better." Make "everything" list its components and show up as cash.

Valuation as a strategy test: value drivers are not abstract

A useful valuation doesn't just answer "how much." It answers "why." Most of that "why" can be traced to a handful of value drivers:

- returns on invested capital (ROIC),
- growth rate,
- and how much reinvestment is required to support that growth.

This is where valuation becomes strategic finance rather than mechanical finance. Two companies can grow at 15% per year. One can create enormous value because it grows with high returns and modest reinvestment. The other can destroy value because it grows by pouring capital into low-return projects. The difference is not the revenue line; it's the economics of reinvestment.

A CFO can use this to elevate conversations with operators. Instead of debating whether a project "increases EBITDA," ask: *does this increase ROIC by improving margins, reducing capital intensity, or increas-*

ing asset turns? That question forces initiatives to justify themselves as value creation, not just activity.

A practical workflow: how to run valuation without turning it into religion

A sane valuation workflow looks like this:

1. Build a driver-based DCF that an operator can understand (units, pricing, churn, utilization, conversion, capex per unit).
2. Cross-check with trading multiples using a carefully chosen peer set and normalized metrics.
3. Cross-check with transaction multiples, adjusting for cycle conditions and deal context.
4. Reconcile differences by identifying the assumptions that would need to change for the methods to converge.
5. Decide on a value range and a walk-away price, not a single point estimate.

The discipline is in step 4. That's where you stop arguing about "the number" and start arguing about the business. If comps imply a higher value than your DCF, is the market expecting higher growth, lower risk, or a scarcity premium? If your DCF is higher than comps, are you underwriting improvements that peers haven't proven, or are comps temporarily depressed by sentiment?

This workflow also makes negotiation sharper. If you know the walk-away price is tied to a small set of assumptions (say, reaching 25% gross margin within 18 months), you can structure terms around those assumptions: earn-outs, seller notes, contingent consideration, or staged investments. Valuation then becomes not just a verdict, but a deal design tool.

Actionable takeaway: write the valuation memo like a skeptic, not a poet

Before any major capital allocation decision that depends on valuation (acquisitions, divestitures, big equity raises, even internal platform investments), produce a one-page valuation memo with three parts:

- *Range*: DCF range, trading comps range, transaction comps range (no single-number pretending).
- *Load-bearing assumptions*: the two or three drivers that swing value the most (often terminal margin, growth duration, reinvestment intensity, discount rate).
- *Walk-away rule*: the maximum price (or minimum value) you'll accept and the specific assumption you refuse to gamble on.

If the memo can't be written clearly, the valuation isn't ready. If it *can* be written clearly, you've done the real work: you've turned valuation from a spreadsheet performance into a decision discipline.

6.3 The Art of the Deal: M&A Strategy and Diligence

The most dangerous sentence in mergers and acquisitions is usually spoken with a grin: "We're in exclusivity." It sounds like progress. It feels like winning. It's also the moment when smart adults begin to behave like they're already married—defending the relationship, overlooking red flags, and explaining away incompatibilities as "integration opportunities."

A CFO's job in M&A is to keep the company from confusing momentum with merit. Deals are not trophies. They are capital allocations with legal signatures. The cost of getting it wrong is not just a write-down; it's years of distracted leadership, employee churn, operational drag, and the slow reputational leak that comes from being the company that "does deals" instead of the company that executes.

Start with ``why" before ``how much" (and make the why measurable)

M&A is often pitched like a menu: new geography, new customers, new technology, new growth. The problem is that "new" is not a strategy; it's an adjective. The CFO's first contribution is to force the deal thesis into a measurable claim that can be tested.

A strong thesis has a narrow spine. It sounds less like "expand into Europe" and more like: "Our customer base is asking for a bundled solution we cannot currently provide; buying this capability should raise retention by X points and lift gross margin by Y through pricing power and lower churn." Now you have something you can diligence. A vague thesis can't be disproven, which means it also can't be improved.

One practical trick: require the sponsor to write the deal thesis as a set of *if-then statements*. If we buy Company A, then within 12 months we can (a) cross-sell product B to X% of their installed base, (b) reduce duplicate G&A by $Z, and (c) shorten our product roadmap by N quarters. If someone can't write the if-then, they don't yet understand what they're buying.

Deal screening: the discipline of saying ``not this one" quickly

The best deals are killed early. That's not cynicism; it's resource management. Screening is where finance turns from analyst to bouncer: not to be rude, but to protect the room.

Screening works when it is ruthless about a few non-negotiables and flexible about everything else. Non-negotiables are the reasons deals become catastrophic:

- a business model that can't produce cash without constant external funding,
- customer concentration that would keep you awake at night,
- unit economics that only look good with perfect growth assumptions,
- regulatory risk that makes closing uncertain or painfully slow,
- cultural mismatch that guarantees executive turnover post-close.

Flexibility is reserved for fixable issues: messy reporting, immature FP&A, lack of procurement discipline, underpriced SKUs. Fixable issues are the ones you can remedy with management and systems. Non-negotiables are the ones that fight back.

A CFO can formalize screening with a one-page intake memo and a quick scoring rubric, but the real value is behavioral: you're teaching the organization that it's normal to walk away. When "walk away" is culturally acceptable, diligence becomes honest. When it isn't, diligence becomes an elaborate justification process.

Valuation is not the thesis; it's the audit of the thesis

A good valuation doesn't create a deal. It interrogates one. Numbers don't make the decision; they expose what the decision depends on.

If the purchase price only works in a rosy case, you don't have a deal—you have a hope. If the economics work even under a sober case, you may have something rarer: a deal with forgiveness baked in.

The CFO's job is to translate the thesis into value drivers and then pressure-test them. If the thesis is cross-sell, then diligence should obsess over customer overlap, sales-cycle reality, quota capacity, and whether bundling actually wins deals. If the thesis is cost synergy, then diligence should detail the org charts, the contracts, the systems overlap, and the one-time costs to remove duplication. If the thesis is "technology," then diligence should investigate not just the codebase, but the team that wrote it and whether they will stay.

That's why the CFO should be allergic to generic synergy slides. Synergy is not a category; it's a list. Every synergy should have an owner, a timing, a cost-to-achieve, and a measurement plan. Otherwise it's just the word you use when the price is too high.

Deal fever: the psychology that turns smart teams into bad buyers

Deal fever is a real phenomenon: as a team invests time, status, and identity into a deal, the desire to close increases regardless of fundamentals. The organization starts optimizing for completion rather than correctness. The sunk costs become emotional instead of financial: "We've already put six weeks into this," "The CEO is excited," "The board wants growth."

A CFO can't eliminate deal fever, but you can design around it. A few mechanisms help:

- *Pre-commit walk-away conditions*: define, early, what findings will kill the deal (e.g., churn above X, revenue recognition issues, gross margin below Y, regulatory timeline beyond Z months).
- *Independent red team*: assign a small group to argue against the deal thesis and present directly to decision-makers.
- *Stage-gated spending*: limit diligence spend until critical risks are cleared; don't fund a full parade until you've checked the bridge.

None of this is about pessimism. It's about preserving the company's ability to say "no" even when the deal starts to feel inevitable.

Due diligence is not a checklist; it's an investigation

Bad diligence asks, "Is anything wrong?" Good diligence asks, "Where is the truth different from the story?" The difference is subtle and everything.

A CFO should treat diligence like building a courtroom case: evidence, corroboration, and a clear theory of what must be true for the deal to succeed. Documents are not enough. Interviews are not enough. Management presentations are definitely not enough. You want triangulation: financials that tie, operational metrics that reconcile, and customer reality that matches both.

If you only remember one principle, make it this: *the seller is not lying to you; the seller is selling to you.* You are responsible for translating salesmanship into economics.

Quality of Earnings (QoE): the CFO's flashlight in a dark room

Quality of Earnings is often misunderstood as "audit-like." It's not. Audit asks, "Is this reported according to accounting rules?" QoE asks, "How repeatable is this profit, and what cash does it really turn into?" That makes QoE the single most valuable diligence workstream for a CFO, because it turns headline EBITDA into something you can underwrite.

A strong QoE typically does four things.

1) Normalizes EBITDA into run-rate earnings

Most businesses have noise: one-time legal settlements, unusual bonus accruals, non-recurring implementation revenue, temporary cost cuts, founder salary quirks. Sellers love to call costs "one-time" and revenues "recurring." QoE tests those claims.

The point is not to be cynical. It's to protect your model from fake smoothness. If the business has $5M of "add-backs," you don't argue about all $5M at once. You categorize:

- *Clearly non-recurring*: plausible add-backs (e.g., a one-off facility repair).
- *Recurring in disguise*: costs that will continue (e.g., "temporary" consultants that have been there for two years).
- *Commercial concessions*: price discounts or service credits that are functionally part of cost of sales.

Every dollar of adjusted EBITDA you accept is a dollar you're paying a multiple on. If you accept shaky add-backs, you are literally paying for them.

2) Tests revenue quality: recognition, concentration, and churn

Revenue can be legitimate and still be fragile. A CFO cares about fragility because fragility drives risk, and risk should change both price and structure.

In diligence, revenue questions should feel uncomfortably specific:

- Are there side letters that change terms?
- Are renewals "automatic" or "auto-billed" (not the same)?
- Are top customers profitable, or are they loss leaders propping up scale?
- Is churn hidden by new bookings (a treadmill business)?
- Are there channel partners whose incentives can flip overnight?

Customer concentration is a special kind of risk because it is both measurable and political. A business can look strong until you notice that one customer is 18% of revenue and gets custom pricing, custom features, and custom support. That's not a customer; that's a co-dependent relationship.

3) Reconciles earnings to cash (working capital and CapEx reality)

A deal can look cheap on EBITDA and expensive on cash. QoE should show how earnings translate into operating cash flow and what reinvestment is required to sustain the business.

Watch for two quiet value killers:

- *Working capital drag*: growth that consumes cash through receivables, inventory, or deferred revenue dynamics you don't fully understand.
- *Understated maintenance CapEx*: a business that has been starved of investment to maximize sale-period EBITDA.

A classic seller play is to cut discretionary spend before selling. Sometimes that's real efficiency. Sometimes it's value extraction. QoE helps you tell the difference by looking for deferred maintenance, rising customer complaints, increased downtime, or sales productivity that "mysteriously" improves while pipeline quality deteriorates.

4) Surfaces liabilities that behave like debt

Not all obligations sit in net debt. Some hide in contracts and habits: deferred revenue performance obligations, warranty exposure, returns reserves, litigation risk, customer rebates, unpaid taxes, or aggressive accrual practices. QoE is where you turn those into quantified issues that can be priced, escrowed, or structured around.

The CFO lens here is simple: *what future cash outflows are already baked in, even if they're not called "debt"?*

Regulatory feasibility: antitrust is now a core finance question

There was a time when regulatory review was treated like a legal footnote: "Our counsel isn't concerned." That era is over for many industries and deal types. Antitrust scrutiny has become a first-order diligence item in the U.S., and CFOs ignore it at their peril.

Regulatory risk changes three things immediately:

- *Timing*: longer review means longer uncertainty, which affects employee retention and customer behavior.
- *Deal certainty*: a non-trivial chance the deal is blocked changes the expected value.
- *Concessions*: remedies (divestitures, behavioral commitments) can carve out the very assets you thought you were buying.

Finance should be in the room for the regulatory strategy conversation because the output is not just a memo; it's a change to the model. If closing shifts from three months to twelve, your synergy timing shifts, your integration plan shifts, and your cost of carry shifts. You may also need a different structure: reverse termination fees, ticking fees, or other provisions that allocate regulatory risk between buyer and seller.

Purchase accounting: why GAAP outcomes matter even when cash is king

It's fashionable to say, "Accounting doesn't matter; we manage cash." That's half true and fully dangerous.

Purchase accounting under U.S. GAAP (ASC 805) requires identifiable assets and liabilities to be recorded at fair value as of the acquisition date. Goodwill becomes the residual. That accounting has real consequences:

- *Amortization of intangibles* can depress future GAAP earnings even when cash is healthy.
- *Deferred revenue fair value adjustments* can reduce post-close reported revenue in ways that surprise sales leadership and spook boards.

- *Impairment risk* creates future earnings volatility if the acquired business underperforms.

A CFO doesn't need to become a technical accountant to manage this. You need to anticipate the narrative. If you know GAAP EPS will dip due to intangible amortization, you prepare stakeholders early and ensure performance scorecards separate operating reality from acquisition accounting artifacts. Surprises are what turn an economically decent deal into a reputational problem.

Deal structure: price is what you pay; terms are how you protect yourself

Two deals with the same headline price can have completely different risk profiles. Terms are where the CFO earns their keep.

A few structural tools are especially CFO-relevant:

- *Working capital mechanisms*: a normalized working capital target prevents the seller from draining the bathtub before handing you the keys.
- *Escrows and indemnities*: protection against unknown liabilities; expensive if you discover them late.
- *Earn-outs*: useful when there's genuine uncertainty about growth, but they can also poison post-close culture if poorly designed.
- *Rollover equity*: keeps sellers invested and can reduce cash outlay, but it changes governance dynamics.

These are not legal decorations. They are financial instruments that allocate risk. A CFO should treat them the way you treat debt covenants: not as boilerplate, but as levers that change expected value.

Operating diligence: the unglamorous facts that decide whether synergies exist

Financial diligence tells you what happened. Operating diligence tells you what *will* happen if you try to combine two living organizations.

This is where M&A gets oddly physical. You're buying warehouses, call centers, production lines, data pipelines, account managers, and routines. Synergies aren't created by spreadsheets; they're created by people changing their work.

A CFO should insist on evidence for operational claims:

- If the thesis is procurement leverage, show vendor overlap, contract terms, and switching costs.
- If the thesis is manufacturing optimization, show capacity constraints, yield, downtime, and quality metrics.
- If the thesis is software platform integration, show architecture, technical debt, release cadence, and the dependency map that will bite you at 2 a.m.

One of the best diligence questions is deceptively simple: *what breaks on day one?* Not quarter one. Day one. If the acquired business relies on the founder to close deals, or on a single engineer to keep systems running, you don't have a scalable asset. You have a fragile organism.

The CFO's deal dashboard: what you track before signing

Before a deal goes to final approval, a CFO should be able to answer these questions without reaching for a slide:

- What is the deal thesis in two sentences, and which metrics will prove it right or wrong?
- What are the three biggest downside risks, and what evidence did diligence uncover about each?

- What is normalized EBITDA and normalized free cash flow, and how did we get there?
- What is the walk-away price and the specific assumption that sets it?
- What is the regulatory path, timeline, and the cost of delay?
- What are the accounting outcomes that will surprise non-finance stakeholders?
- What terms protect us if the story is wrong?

If these answers are fuzzy, the deal is not ready. If they are crisp, you have something rarer than confidence: you have *earned* conviction.

Actionable takeaway: institute a ``Thesis-to-Test" memo that makes diligence concrete

Adopt a mandatory one-page "Thesis-to-Test" memo for every serious target before exclusivity:

- *Thesis*: the if-then statements (measurable, time-bound).
- *Tests*: the exact diligence workstreams that will validate or falsify each claim (QoE items, customer calls, cohort analyses, contract reviews, operational KPIs).
- *Walk-away triggers*: three specific findings that kill the deal, pre-approved by the decision group.

This memo changes behavior immediately. It turns diligence from a frantic document hunt into an investigation with a purpose. It also gives the CFO the cleanest form of authority in M&A: not the authority to say "no," but the authority to say, "Prove it."

6.4 After the Handshake: Realizing Deal Value

The day the deal closes, the spreadsheet becomes historical fiction.

Until closing, everyone is paid to imagine. Bankers imagine a clean story. Executives imagine a bigger platform. The model imagines synergies arriving on schedule like polite guests. Then the wire goes out, the press release posts, and the organization wakes up to a less cinematic truth: two companies now have to share payroll calendars, customer promises, data definitions, and a thousand tiny habits that used to be someone else's problem.

That's why so many deals "fail" in a way that's maddeningly quiet. They don't explode. They leak. Revenue drifts because sales teams hesitate. Costs rise because integration takes longer than the model admitted. Good people leave because the new org chart makes them feel small. Customers notice friction long before the CFO sees it in the P&L. The wedding photo looks great; the marriage is fought in the unglamorous rooms: billing, IT access, customer support scripts, purchase orders, and who is allowed to approve a discount.

A CFO's role after close is simple to state and hard to execute: *turn promised value into bankable cash while protecting the base business*. That means two simultaneous jobs that naturally conflict. One is value capture (synergies, growth, efficiency). The other is business continuity (ship, invoice, collect, comply). The CFO becomes the person who refuses to let synergy ambition break Day 1 reality.

Day 1 is a cash flow event, not a branding event

Integration teams love Day 1 decks: new logos, new leadership messages, new mission statements. Day 1 is actually something more fragile: the first day your combined company tries to perform the basic motions that create cash.

If invoicing fails, revenue becomes an argument. If collections stall, the cash conversion cycle stretches immediately. If shipping is disrupted, customers defect and you learn about it later through lagging churn metrics. If payroll hiccups, trust collapses in a single afternoon.

Treat Day 1 readiness like a going-concern exercise. The question is not "Are we integrated?" The question is "Can we operate without breaking cash?" A CFO who frames it this way changes priorities instantly. Suddenly, the most important integration meeting is not the synergy brainstorm; it's the one where someone explains how customer master data will be migrated without corrupting credit terms.

Here's the CFO-grade Day 1 mindset: *stabilize first, optimize second.* Optimization is where the synergy lives. Stabilization is where the deal survives long enough to see it.

Synergies: the most abused word in corporate language

Synergy is supposed to mean that together you can produce more value than separately. In practice, synergy often means, "We needed a reason the price made sense."

The cure is to stop talking about synergy as a lump sum and start treating it as a portfolio of initiatives. Each initiative needs four attributes to be real:

- *Mechanism*: what operational change creates the benefit?
- *Timing*: when does the benefit start, and when does it reach full run-rate?
- *Cost to achieve*: severance, systems work, consultants, retention bonuses, duplicate run costs.
- *Measurement*: the metric that proves the benefit is showing up in reality, not just in a PowerPoint.

If any one of these is missing, you don't have a synergy. You have a hope dressed up as arithmetic.

Synergies also come in two flavors with very different personalities:

- *Cost synergies* are usually more controllable (fewer vendors, fewer duplicate roles), but politically painful and operationally risky if rushed.
- *Revenue synergies* are usually more exciting (cross-sell, bundling, channel expansion), but less controllable and slower, because they depend on customers changing behavior.

A CFO who treats revenue synergies as "upside" rather than "required" is not being conservative. You're being accurate about human behavior.

The first 100 days: where momentum is made or squandered

There's a reason experienced operators care obsessively about the first few months after close: the organization is unusually malleable, and attention is unusually concentrated. Momentum is available like a limited-time discount. If you don't spend it on the right things, it expires.

The CFO's most important decision in the first 100 days is not a model tweak; it's governance design. Who owns integration? Who decides trade-offs between speed and safety? What gets escalated, and how quickly? Which leaders are measured on integration outcomes, and what consequences are real?

The wrong governance produces a predictable pathology: everything becomes "urgent," so nothing is prioritized, so execution stalls, so the organization drifts back to old habits, now with more meetings.

The right governance feels almost boring. Clear workstreams. Clear owners. Clear decision rights. A weekly cadence that surfaces issues early and forces resolution. Boring is good. Boring means you're integrating a company, not improvising a reality show.

Integration is not one project; it's a stack of dependencies

The biggest lie of many post-merger plans is the Gantt chart that implies tasks are independent. In reality, integration is a dependency stack. IT access affects finance close. Finance close affects performance reporting. Performance reporting affects incentive comp. Incentive comp affects sales behavior. Sales behavior affects revenue. Revenue affects morale, because morale follows momentum.

This is why integration is often gated by IT even when the deal thesis has nothing to do with technology. Systems are how the business thinks and remembers. If the systems can't talk, the organization becomes bilingual in the worst way: two definitions of revenue, two customer IDs, two ways to approve spend. Confusion is expensive.

A CFO doesn't need to be the CIO to manage this. You need to insist on one artifact: an integration blueprint that shows which systems will be the "survivor" for each major process (order-to-cash, procure-to-pay, record-to-report, hire-to-retire), the timeline for cutover, and the risks to continuity. If you don't pick survivors early, you'll end up with a hybrid creature that costs more than either original company and is loved by no one.

Synergy tracking: separate the math from the mood

The fastest way to lose synergy value is to let it dissolve into business-as-usual. The base business grows and shrinks for many reasons, and if you don't isolate synergy initiatives, you'll never know whether you captured value or merely got lucky with the market.

A CFO should set up synergy tracking as a parallel ledger of accountability:

- Every synergy initiative has an ID, an owner, and a baseline.
- Benefits are tracked at the same cadence as financial reporting.

- One-time costs are tracked alongside run-rate savings, so the payback story stays honest.
- Benefits are mapped to financial statement lines, so the P&L eventually corroborates the narrative.

The mapping part matters because integration teams often claim wins that don't hit the financials. "We consolidated vendors" is not a win. "Cost of goods sold decreased by $X with no service degradation" is a win. "We reduced headcount" is not a win if contractors quietly replace those roles three months later. The CFO's job is to prevent substitution tricks, where the organization meets a target by changing labels instead of economics.

A useful concept here is *leakage*: the gap between identified synergy and realized financial impact. Leakage happens because savings are re-spent, because costs move to different buckets, because managers defend budgets, or because operational disruption creates offsetting costs. If you don't explicitly measure leakage, you will consistently overestimate integration success.

The valley of despair: why synergies often get worse before they get better

Integration has an unromantic physics: combining two systems usually creates friction before it creates efficiency. You carry duplicate costs while you migrate. You pay severance before you see savings. You spend on systems before you can retire legacy platforms. You distract top performers with integration work, then wonder why performance dips.

This dip is not a sign the deal is bad. It's a sign the integration is real.

The CFO's role is to make the valley explicit in the plan so it doesn't become a surprise accusation later. That means budgeting integration costs as a program, not hiding them in departmental budgets where they'll be resented and underfunded. It means communicating to the board and leadership team that synergy is a *curve*, not a light switch. And it means protecting the base business with capacity planning, because the base business is the thing paying for the integration.

A helpful tactic is to separate reporting into two views:

- *Run the business*: the operating P&L and KPIs without integration adjustments.
- *Run the integration*: synergy progress, integration costs, and risks tracked like a transformation program.

That separation keeps leaders from making panicked cuts when integration costs rise and keeps the integration team from declaring victory because the economy improved.

People risk is financial risk wearing a human face

Culture is often treated like soft stuff: posters, values, town halls. Culture becomes brutally quantitative in M&A. If key talent leaves, product roadmaps slip, customer relationships weaken, and execution risk rises. None of that shows up immediately as a line item called "culture." It shows up as missed numbers and then, later, as goodwill impairment conversations nobody enjoys.

CFOs can manage people risk without pretending to be HR. The trick is to treat talent as a value driver with leading indicators:

- retention rates in critical roles (sales leaders, engineers, plant managers),
- regretted attrition,
- time-to-fill for backfills,
- employee engagement pulse surveys focused on clarity and trust,
- and customer-facing staffing stability.

Then connect those to the deal thesis. If the deal depends on cross-sell, sales leadership stability is not "nice." It's structural. If the deal depends on a technology platform, engineering retention is not "HR's problem." It's part of the purchase price justification.

When finance tracks these indicators alongside synergy metrics, the organization stops treating talent loss as bad luck and starts treating it as managed risk.

The CFO integration tool kit: a few levers that actually move outcomes

Post-close, the CFO's power is less about authority and more about instrumentation. You build the dashboard the company drives by. Choose it well and behavior follows.

A high-functioning finance-led integration typically includes:

- *A single source of truth for synergy numbers*: one model, one owner, one version.
- *A monthly synergy close*: benefits and costs booked or reconciled like financials, not casually updated.
- *A benefits "burndown"*: planned vs. actual run-rate achieved over time, with explanations that aren't vague.
- *A working capital watchtower*: immediate focus on invoicing, collections, credit terms, and customer disputes during systems changes.
- *Integration cost governance*: approval gates for integration spend so it doesn't balloon under the banner of "necessary."

One underappreciated lever is incentive design. If leaders are paid on combined EBITDA without adjusting for integration spend, they will starve integration. If they are paid on synergy promises without measuring base business health, they will break continuity. Balanced incentives are not a philosophy; they are an operating requirement.

When to admit the plan is wrong (and why that's not failure)

Integration plans are made under uncertainty. The sin is not being wrong. The sin is clinging to the plan to protect pride.

A CFO should build explicit re-forecast points into the integration governance: moments when the organization revisits synergy timing, costs-to-achieve, and the operational reality without shame. Treat it like a rolling forecast. If IT migration slips by three months, it's better to admit it, update the synergy curve, and protect customer experience than to pretend the original timeline still exists.

In fact, one of the most valuable contributions finance can make is reframing "missed synergy timing" as a decision trade-off rather than a moral failure. Sometimes it's correct to slow down to protect customers. Sometimes it's correct to accelerate and accept disruption. What's unacceptable is drifting because nobody wants to be the messenger.

Actionable takeaway: run a ``Synergy Close" like a mini-financial close

Starting the first full month after closing, implement a monthly *Synergy Close* with the same seriousness as your financial close:

- lock synergy definitions (what counts, what doesn't),
- update actuals for both benefits and costs-to-achieve,
- reconcile to the P&L and cash flow where possible,
- and publish a short variance commentary that names owners and next actions.

This one habit prevents the most common post-merger failure mode: synergies that look real in meetings but never quite become money. When synergy becomes a close process rather than a slide, the deal's value stops being theoretical and starts showing up where it matters—in cash, in margins, and in credibility.

Chapter 7

The Shield: Risk, Governance, and Tax

Most people think of the CFO as the scorekeeper. But when the storm hits-whether it's a sudden currency crash, a lawsuit, or a surprise tax audit-the CFO becomes the shield. This chapter moves beyond checking boxes to building a defense system that ensures your company survives the bad days so it can thrive on the good ones. We will turn terrifying concepts like 'derivatives' and 'governance' into practical armor you can wear every day, ensuring that when risks materialize, your organization doesn't just panic-it pivots.

7.1 Enterprise Risk Management: Building the Radar

At 2:13 a.m., nobody cares that your quarter closed on time.

The CEO cares that your biggest customer just froze payments because their bank flagged a sanctions issue. Sales cares because the renewal was supposed to fund next month's hiring plan. Engineering cares because the customer's data pipeline is now paused and your SLA clock is still ticking. You care because the cash forecast you signed off on six days ago is now fiction—and every decision that depends on it (inventory, headcount, debt covenants, investor messaging) starts wobbling.

That moment is when most companies *discover* their risk program. Good ones already had it running.

Enterprise Risk Management (ERM) is not a binder. It's a radar: a living system that continuously scans for threats and opportunities, translates them into business language, and forces decisions while there's still time to choose. A risk that shows up early is a management problem. A risk that shows up late is a crisis, and crises are the most expensive way to learn.

Risk is not bad news; it's a price tag on your strategy

Businesses don't fail because they take risk. They fail because they take risk *blindly*-or because they take risks they didn't mean to take while obsessing over the ones they did.

Launching a new product is risk. Extending payment terms to win a large customer is risk. Concentrating production in a single region because it's cheaper is risk. Even choosing *not* to act is risk: the risk of being outpaced, commoditized, regulated out of a market, or simply forgotten.

ERM works when it does two things at once:

- It protects the downside (preventing the surprise that breaks you).
- It enables the upside (making bold moves with your eyes open and your escape routes planned).

The CFO is the natural operator of this system because finance is where risks show up with their masks off. Operational problems become margin compression. Vendor failures become working capital spikes. Legal disputes become contingent liabilities. Cyber incidents become revenue interruptions and disclosure clocks. Risk is just uncertainty that hasn't been priced correctly yet.

Start with a plain-language definition: what counts as a risk here?

If you ask ten executives what "risk" means, you'll get ten answers and at least three therapy sessions. ERM begins by forcing a shared definition:

A risk is any uncertainty that could materially affect your ability to hit objectives-positively or negatively.

That "positively" matters. A sudden competitor exit is an upside risk (a chance to gain share fast) that still carries execution and reputational hazards if you over-promise. A favorable regulatory change is upside risk that can pull you into a market before you've built the muscle to serve it. ERM stops being a compliance chore when it becomes the place where strategic surprises-good and bad-get handled with discipline.

A practical way to keep the definition from drifting into philosophy is to anchor it to objectives that already exist:

- *Cash objectives:* maintain a minimum liquidity buffer; avoid covenant breaches; fund payroll and critical suppliers.
- *Earnings objectives:* protect gross margin bands; manage volatility; avoid "one bad quarter" narratives.
- *Operational objectives:* meet SLAs; maintain quality; keep critical systems within recovery targets.
- *Strategic objectives:* expand markets; reduce concentration; win regulated customers; pass diligence.

If a risk cannot be tied to an objective, it's either noise-or you haven't articulated your objectives clearly enough.

Risk appetite: the rules of the road, not a motivational poster

"Risk appetite" is the amount of risk the company is willing to accept in pursuit of its goals. Most companies write something vague like "we have a moderate risk appetite," which is roughly as useful as telling a pilot you have a "moderate appetite for turbulence."

The CFO's job is to help turn appetite into *decision rules*.

Think of risk appetite as guardrails:

- Guardrails are not the same as a plan. They define where driving becomes reckless.
- Guardrails are not the same as limits. Limits are measurable thresholds that trip action.
- Guardrails are not "no." They're "yes, if," and the "if" is what keeps you alive.

A good risk appetite statement has two layers:

Layer 1: Principle-level appetite (*human-readable*). Example: "We take product and market risk to grow, but we avoid risks that could threaten liquidity or regulatory standing."

Layer 2: Operational appetite (*measurable triggers*). Examples that actually influence behavior:

- *Liquidity:* "We will not operate with less than X months of cash runway under the downside case."
- *Customer concentration:* "No single customer should exceed Y% of revenue without board-level review and a contingency plan."
- *Pricing/margin:* "We accept promotional margin compression up to Z points for up to N quarters, but not structural margin degradation without a path back."
- *Security/availability:* "Critical systems must meet recovery time objectives; deviations require formal risk acceptance with an owner and expiration date."

The most powerful phrase in ERM is *formal risk acceptance*. It means: "We see the hazard, we choose to live with it, we name the owner, and we set the conditions under which we'll revisit the choice." That single habit prevents the most common corporate lie: "We didn't know."

Build a risk taxonomy so people stop arguing about words

When Marketing says "risk," they might mean brand damage. When Engineering says it, they mean uptime and technical debt. When Legal says it, they mean liability. When Finance says it, they mean cash and covenants. Everyone is right-and the company still loses because nobody is speaking the same language.

A risk taxonomy is simply a shared menu of risk categories. It does not need to be academic. It needs to be usable in meetings.

A practical taxonomy for most operating companies:

- *Strategic:* market shifts, competitor moves, product bets, M&A integration.
- *Financial:* liquidity, interest rates, FX, credit risk, covenant risk, fraud.
- *Operational:* supply chain, quality, capacity, vendor reliability, process breakdowns.
- *Technology & cyber:* outages, incidents, access controls, data integrity, third-party risk.
- *Legal & regulatory:* compliance, contracting, IP, privacy, sector regulations.
- *People:* key-person dependency, hiring constraints, culture, safety, labor issues.
- *Reputational:* brand trust, customer harm, public scrutiny, social media volatility.

The taxonomy's hidden benefit is political: it prevents a single function from becoming the "risk police." When risks are categorized, they naturally find owners-because categories map to operating responsibilities.

From brainstorm to discipline: identifying risks without creating a panic festival

Risk identification fails in two classic ways:

- *The theater problem:* the team generates an impressive list, files it, and nothing changes.
- *The anxiety problem:* the list becomes a dumping ground for every fear, and leadership stops listening.

The fix is to use multiple lenses and insist on specificity.

Useful lenses that keep risk identification grounded:

- *Objective lens:* "What would stop us from hitting the plan, the cash forecast, or the customer commitments?"
- *Process lens:* "Where do handoffs fail? Where does a single spreadsheet cell decide the truth?"
- *Concentration lens:* "What are we over-dependent on-customers, suppliers, systems, people, geographies?"
- *Change lens:* "What's changing-regulation, technology, pricing models, buyer behavior?"
- *Stress lens:* "What breaks under load-end of quarter, peak season, a sudden demand spike?"

Then push each risk statement to include four elements:

- *Trigger:* what event starts the chain?
- *Pathway:* how does it propagate through operations?
- *Impact:* what does it do to cash, earnings, customers, or compliance?
- *Time-to-pain:* how quickly do you feel it?

Time-to-pain is the CFO's secret weapon. A risk that can hurt you in 48 hours (payroll system outage, ransomware, key supplier failure) demands different controls than one that plays out over 18 months (slow commoditization, shifting buyer preferences). ERM becomes practical when it respects tempo.

Prioritization: the moment you stop pretending you can fix everything

Once risks are identified, prioritization is where maturity shows. If everything is "critical," nothing is. If the list is ranked by whoever speaks loudest, it's not ERM-it's improv.

Most companies prioritize using two dimensions:

- *Probability:* how likely is it?
- *Impact:* how bad is it if it happens?

But the trick is choosing what "impact" means. For CFO-led ERM, impact should be defined in terms executives can't ignore: liquidity, profitability, compliance exposure, customer harm, and strategic derailment. When impact is translated into those currencies, arguments become decisions.

A clean way to visualize this is a 2x2 risk heat map. It's intentionally blunt-like a triage board in an ER. The goal is not precision; it's prioritization that drives action.

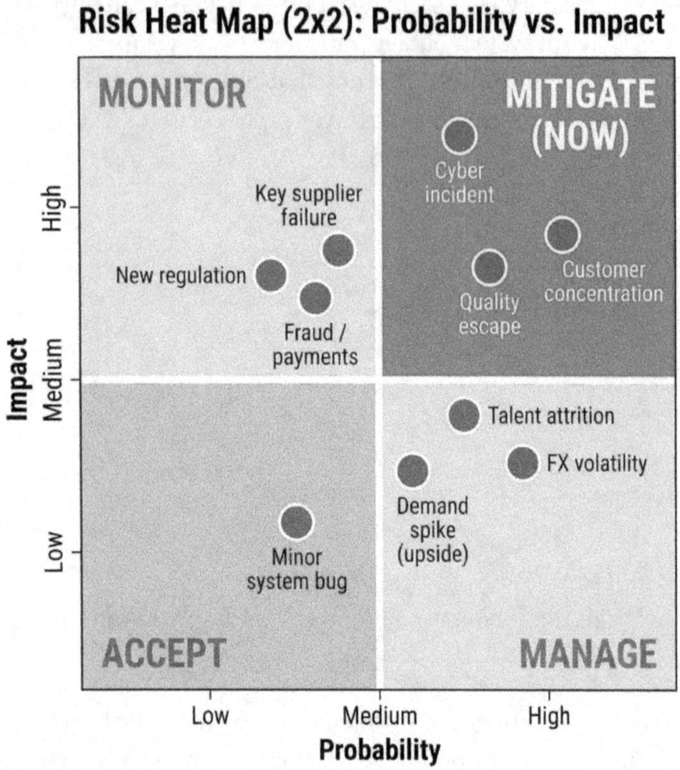

A heat map is only honest if the scoring is honest. Two practices keep it from becoming corporate art:

- *Calibrate probability.* Use ranges ("unlikely," "possible," "likely") tied to historical frequency or comparable events. If nobody can articulate what would make a risk "likely," the score is a guess wearing a tie.
- *Define impact thresholds in dollars and days.* "High impact" might mean "more than $X cash outflow," "more than Y days of revenue disruption," "a covenant breach," or "a reportable incident." When thresholds are explicit, scoring becomes faster and less political.

Treatment options: four verbs that stop endless debate

Once a risk is prioritized, treatment should collapse into four verbs:

- *Avoid:* stop doing the activity that creates the risk (exit a market, stop selling a product with unacceptable liability).
- *Mitigate:* reduce probability or impact (controls, redundancy, process redesign, contractual protections).
- *Transfer:* shift financial exposure (insurance, indemnities, hedging, outsourcing with strong SLAs).
- *Accept:* live with it intentionally (with an owner, monitoring, and a trigger point).

The CFO can make this brutally practical by asking one question that turns philosophy into arithmetic:

What's the cheapest reliable way to move this risk out of the red quadrant?

"Cheapest" is not stingy. It's strategic. Over-controlling low-impact risks is how companies quietly tax themselves into mediocrity. Under-controlling high-impact risks is how they end up explaining themselves to a board-or a court.

Governance: who owns the risk when nobody wants to raise their hand?

ERM fails most often because ownership is fuzzy. When a risk spans functions (most do), it becomes an orphan.

A working governance model has three layers:

1) *The Board (or its delegate) sets the appetite.* They don't manage the risks day-to-day, but they set the boundaries: how much volatility is acceptable, what existential risks are off-limits, and what must be escalated.

2) *Management owns risks and runs responses.* Every top risk needs a single executive owner. Not a committee. Not "shared." One name.

Ownership means being accountable for mitigation plans, timelines, and residual risk (the risk left after controls).

3) *Finance/Risk orchestrates the system.* This is the CFO's leverage point: not "owning all risks," but owning the *process* that forces visibility, prioritization, and follow-through. That includes scoring standards, reporting cadence, and linking risk discussions to planning and performance.

The best ERM meeting is boring in a specific way: no surprises, no melodrama, and no vague commitments. Every high risk has a plan, a date, and a measurable way to tell whether it's improving.

Tie ERM to planning, or it becomes a hobby

The moment ERM becomes real is when it changes one of these:

- the budget,
- the hiring plan,
- the capital plan,
- customer and vendor terms,
- or the way performance is measured.

If your risk register says "key supplier dependency" is a top risk but procurement is still measured only on unit cost, the organization has voted. If the register says "cash volatility" is a top risk but Sales is rewarded for bookings with no regard to collectability, the organization has voted.

A CFO can hardwire ERM into operations by linking it to three rhythms that already exist:

Forecasting rhythm. For the top risks, define how they would show up in the forecast (revenue delay, DSO expansion, gross margin hit, inventory write-down, emergency capex). Then track leading indicators alongside financial outcomes. The forecast becomes not just "what we think will happen," but "what would need to break for this to be wrong."

Operating cadence. Top risks should appear in leadership dashboards with a small set of indicators (not twenty). If it's truly top-five, it deserves ongoing airtime.

Decision memos. Any decision that materially changes exposure-large contract terms, new geography, major system change-should include a paragraph: what risks increase, what risks decrease, and what mitigations are being funded. This is how ERM becomes muscle memory instead of a quarterly ritual.

Make it measurable: leading indicators beat autopsies

A risk program that only reports incidents is a history club. The point is to see problems early.

Leading indicators are signals that move *before* the bad outcome:

- *Customer concentration risk:* pipeline diversity, renewal health scores, top-customer AR aging.
- *Cyber/tech risk:* patch latency, privileged access sprawl, backup restore test success rate, third-party security posture.
- *Liquidity risk:* daily cash variance vs forecast, borrowing base headroom, collections performance by cohort.
- *Supply chain risk:* on-time delivery rates for critical vendors, single-source components count, lead-time volatility.

The CFO's advantage is that finance already runs an early-warning network: variances, collections, churn cohorts, margin bridges, and cash burn. ERM should harvest those signals and route them to the people who can act, not just the people who can report.

A short case: the risk you didn't mean to take

Consider a company that prides itself on being "capital efficient." It negotiates longer payment terms with suppliers to preserve cash. It also offers generous terms to win enterprise customers. Separately, each decision sounds reasonable. Together, they can create a quiet liquidity trap.

If customer payments slip by 20 days in a quarter, and supplier payments are already stretched, the company has no flex left. Suddenly, the "capital efficient" strategy becomes "we are one collections miss away from a scramble." That scramble has a predictable shape: emergency borrowing at ugly rates, delaying payroll-related investments, pressuring teams to ship risky releases to pull revenue forward, and a spike in internal friction that makes the next quarter worse.

ERM would label this as a cross-functional risk (commercial terms + working capital + vendor relations) with a clear owner, measurable thresholds, and pre-planned actions:

- Avoid/mitigate by tightening deal desk terms when concentration is high.
- Transfer by using credit insurance selectively or structuring milestone billing.
- Accept only with defined liquidity buffers and a trigger point that forces action *before* cash gets tight.

No heroics. Just design.

Actionable takeaway: run a 30-day ERM sprint that produces decisions, not documents

A month is enough time to build a radar that actually works-if the goal is *operating clarity*, not perfection.

1. *Write three appetite guardrails* in plain language and attach one metric to each (liquidity, concentration, compliance/availability are usually the fastest wins).
2. *Build a one-page risk taxonomy* and require every new risk to be categorized and assigned an owner.
3. *Identify the top 10 risks* using triggers, pathways, impact, and time-to-pain; score them on probability/impact with explicit thresholds.
4. *For the top 5, choose a verb* (avoid/mitigate/transfer/accept) and fund one concrete mitigation per risk with a date.
5. *Attach one leading indicator* to each top risk and put it into the same dashboard cadence as cash and performance metrics.

If the sprint ends with five named owners, five funded actions, and five indicators that show up in real meetings, you have an ERM program. If it ends with a beautiful spreadsheet and no behavior change, you have corporate stationery.

7.2 Financial and Operational Protection: The CFO's Armor

A good CFO has two personalities.

One loves growth: new markets, bigger contracts, bolder pricing, faster hiring. The other is quietly paranoid and sleeps with one eye open. That second personality is the one that asks, at the exact wrong moment, "What happens if the euro drops 8% in a month?" or "What happens if our flagship facility is offline for ten days?" or "What happens if a ransomware crew decides we look like a fast payday?"

The point isn't to be fearful. The point is to make sure your company's performance isn't determined by whatever headline the world decides to print next.

Financial and operational protection is the CFO's armor: a set of instruments, contracts, and insurance decisions that *cap the damage* of volatility and disasters. The goal is not to eliminate risk—that's impossible and, frankly, would kill the business. The goal is to take risks you *intend* to take (product, market, execution) and stop taking risks you didn't mean to take (currency roulette, interest-rate whiplash, one-plant dependency, one-vendor fragility, catastrophic liability).

The CFO's prime directive: trade uncertainty for known costs

Most protective tools feel "expensive" because their benefit is invisible on a good day. A forward contract looks like unnecessary paperwork when FX is calm. Business interruption insurance looks like a waste until the week you need it. A tighter credit policy looks like "slowing sales" until collections freeze and your cash forecast starts lying to you.

The right mental model is simple: protection trades a potentially huge, ugly surprise for a smaller, known cost. That cost can be:

- a premium (insurance),
- a spread or fee (hedging),
- a lower upside (collars, caps),
- or a real operational expense (redundancy, dual sourcing, controls).

If you only buy protection when it feels cheap, you'll buy it after the world has already re-priced it. Protection is best when it's boring—renewed on schedule, tested, documented, and tuned to the business you actually run.

Hedging without mystique: a derivative is just a seatbelt

Derivatives have an undeserved reputation. They sound like casino chips because some companies have used them that way. But at operating companies, the most common derivatives are not speculative. They're seatbelts.

A derivative is a contract whose value is linked to something else: an exchange rate, an interest rate, a commodity price. Hedging means using that contract to reduce the volatility of your cash flows or margins.

The CFO-level question is not "Are derivatives risky?" Everything is risky. The real question is: *Are you currently exposed to a risk that you're not being paid to take?* If the answer is yes, a hedge may be the cheapest way to stop donating money to randomness.

Foreign exchange (FX): the silent margin killer

FX risk is the classic "we'll deal with it later" problem—until later shows up as a gross margin miss that nobody can explain without a geography lesson.

FX exposure comes in two flavors:

- *Transaction exposure:* you will pay or receive foreign currency in the future (e.g., you invoice a European customer in euros but report in USD).
- *Economic exposure:* FX shifts change your competitive position (e.g., your competitor's costs are in a weakening currency, and they can undercut you).

The first is hedgeable with straightforward tools. The second is strategic and usually requires pricing and sourcing decisions—not just treasury.

A simple example makes it real. Suppose you sell a $1,000 product to a customer in France. They pay €920 today. When you set the price, the exchange rate is 1.09 USD/EUR, so you expect roughly $1,003. But you're collecting in 60 days. If the euro weakens to 1.02 by the time you get paid, that same €920 turns into $938. Your "same" sale just lost $65 to currency. No customer negotiated you down. The product didn't change. You just ran an FX trading desk by accident.

FX forwards: locking an exchange rate like locking a mortgage rate

An FX forward is the plainest hedge: you agree today on the exchange rate for a future conversion. It's not sexy, but it's powerful. You're essentially telling the market, "I don't want to guess. Give me certainty."

Mechanically:

- You forecast a foreign-currency inflow or outflow (say €5M of receivables over the next quarter).
- You enter a forward contract to sell euros for dollars at a fixed rate on a set date (or a set window).
- When you collect euros, you convert them at the locked rate via the forward.

What you *give up*: if the euro strengthens, you won't benefit.
What you *gain*: your margin stops being a weather report.

Two CFO rules keep FX hedging from turning into accidental speculation:

- Hedge *exposures*, not opinions. The trigger is forecasted cash flows, not a view on where currencies are going.
- Use a hedge ratio that matches forecast quality. If your forecast is shaky, don't hedge 100% and then panic when volumes miss. Many teams start with 50–80% of high-confidence exposures and scale as forecasting discipline improves.

Options: when you want a floor, not a fixed ceiling

Forwards give certainty. Options give *asymmetry*: protection against bad moves while keeping some upside if moves go your way.

A currency option is the right, not the obligation, to exchange at a set rate. You pay an upfront premium for that flexibility. Options make sense when:

- volumes are uncertain (you might not have the euros to deliver into a forward), or
- you want downside protection but don't want to fully cap upside, or
- the business is making a high-stakes bet (a launch quarter) where bad FX timing could distort the story.

Options are also where treasury can get overly clever. The CFO's discipline is to keep the objective simple: *protect the plan and the cash.* If the strategy needs a spreadsheet to explain why it isn't speculative, it probably is.

Interest rates: the risk hiding inside ``normal" debt

Interest rate risk doesn't announce itself with drama. It just quietly re-prices your cost of capital and then shows up everywhere: in debt service, valuation, customer demand, and covenant headroom.

The exposure is straightforward:

- If you have floating-rate debt, rate increases can raise interest expense fast.
- If you rely on refinancing, rate spikes can close doors right when you need them open.

The most common hedges here are:

- *Interest rate swaps:* exchange floating payments for fixed payments (turning variable interest into predictable interest).
- *Caps:* pay a premium to limit how high the floating rate can go (like insurance on your interest expense).

A practical CFO way to decide is to treat it like budgeting for pain:

- If rates rise to X, what happens to monthly interest expense?
- Does that break covenants? Does it compress runway below your comfort level?
- If yes, what's the cheapest tool to keep the downside inside your risk appetite?

Swaps are often cheaper than caps but remove upside if rates fall. Caps cost upfront but preserve the ability to benefit from lower rates. The "right" answer depends less on macro predictions and more on how fragile your cash flow is to rate moves.

Commodities and energy: hedge the input that can erase your pricing power

For many businesses, commodities are the real FX—materials, fuel, packaging, electricity. If you cannot reprice quickly, commodity volatility becomes a margin trap.

Hedging commodities can get complex quickly (basis risk, contract terms, volume uncertainty). The CFO's job is to keep the scope tight:

- Hedge the portion of costs that are both material and hard to pass through.
- Use simple instruments and short horizons at first.
- Demand a clear link between hedge positions and physical exposure (purchase orders, production plans).

If you can pass through costs quickly via pricing clauses, the best hedge may be contractual rather than financial. If you can't, a limited hedge program can buy you time to adjust pricing without turning earnings calls into a commodity commentary show.

The governance that keeps hedging from becoming gambling

A hedge program isn't defined by its instruments. It's defined by its governance. The instrument is the easy part; the discipline is the hard part.

A hedging policy should answer, in plain language:

- What exposures are eligible (FX receivables, forecasted purchases, floating-rate debt)?
- What instruments are allowed (forwards, swaps, caps; avoid exotic structures unless you have real expertise)?
- What horizon and hedge ratio are permitted?
- Who can approve trades, and what reporting is required?
- What counts as effectiveness (did it reduce volatility of cash flow or margin as intended)?

Then add the cultural rule that saves careers: *treasury does not take views*. If a bank pitches a "smart" structure with a narrative about market direction, you should feel your risk radar start screaming.

Insurance: buying someone else's balance sheet for the worst day of your life

Insurance is the most under-loved part of corporate finance because it feels like paying for nothing. Then a fire, flood, lawsuit, or outage turns "nothing" into "the difference between survival and a funding round at gunpoint."

Insurance is not a substitute for controls, and insurers will punish you for acting as if it is. But it is an efficient way to transfer catastrophic loss—loss that would otherwise overwhelm your liquidity buffer and force strategic concessions.

The CFO should think about insurance as a portfolio, not a shopping list. The question is: *Where could one event create a loss that is too big relative to our cash and earning power?* Those are the events you insure first.

Common corporate coverages (varies by industry) include:

- *General liability* (bodily injury/property damage claims).
- *Product liability* (especially for hardware, consumer, medical, automotive).
- *Professional/E&O liability* (errors and omissions; critical for service and software).
- *Cyber liability* (incident response, legal costs, some business interruption depending on policy).
- *Property* (damage to facilities/equipment).
- *Business interruption* (lost income from covered events—read definitions closely).
- *D&O* (directors and officers; protects leadership and helps recruit it).
- *Workers' comp* and *employment practices liability* (people risks that turn into litigation).

A subtle but crucial point: policies are made of definitions. "Business interruption" can mean wildly different things depending on triggers and exclusions. "Cyber" can cover response costs but not the revenue loss you assumed it would. The CFO's job is to push past the headline name and interrogate coverage mechanics.

Deductibles, retentions, and the art of not over-insuring

Insurance decisions are capital allocation decisions wearing a risk-management costume.

A high deductible (or self-insured retention) means you keep more risk on your balance sheet in exchange for lower premium. That's rational when:

- you can absorb frequent, smaller losses without threatening liquidity, and
- you want insurance mainly for catastrophe.

A low deductible means you pay more to smooth smaller losses. That can be rational when:

- claims frequency is high and disruptive, or
- cash flow is tight and volatility itself is dangerous.

The CFO-level framing is: *How much loss can we self-fund without turning operational problems into financing problems?* Then set deductibles so routine issues don't become insurance paperwork, while true disasters don't become existential.

Operational risk transfer: contracts as armor plating

Some of the best "insurance" isn't bought from an insurer. It's negotiated.

Contracts can transfer or reduce risk through:

- *Indemnities:* who pays if something goes wrong (product defects, IP claims, data breaches).
- *Limitations of liability:* caps on damages; exclusions of consequential damages.
- *Service level agreements (SLAs):* remedies for downtime, performance failures, late delivery.
- *Force majeure and termination clauses:* clarity on what happens when the world breaks.
- *Payment terms and security:* deposits, milestone billing, letters of credit.

A CFO who treats contracting as "Legal's thing" misses a lever that can protect cash more effectively than a hedge. One aggressive customer clause can create a risk exposure larger than your entire insurance program.

The best practice is boring and powerful: require a finance review on non-standard terms that materially affect cash, liability, or performance obligations. Not to slow down deals—but to prevent a single contract from rewriting your risk appetite without permission.

Business continuity: operational resilience is a finance problem

Disaster planning is usually filed under operations or IT. It becomes the CFO's problem the moment you attach numbers to it.

A business continuity plan answers:

- What is the maximum tolerable downtime for each critical process?
- What resources are needed to recover (people, vendors, systems, data)?
- What does the downtime *cost* per day in revenue, margin, penalties, and cash?

When you quantify the cost of downtime, resilience investments stop being "nice to have." They become ROI-positive projects:

- redundancy for a critical system,
- dual sourcing for a single-point-of-failure component,
- backup testing that actually proves recovery,
- alternate logistics routes,
- documented manual workarounds for billing and payroll.

Finance has a unique role here: insisting that continuity plans are tested, not admired. A plan that hasn't been rehearsed is a bedtime story.

The ``unsexy" controls that prevent sexy disasters

Not every protection tool is a financial instrument or a policy. Many are operational controls with outsized impact on loss frequency:

- segregation of duties in payments (so one person can't create and approve a vendor),
- vendor onboarding checks (bank account validation, sanctions screening where relevant),
- approval thresholds tied to risk, not just spend,
- monitoring of unusual transactions and credit memo patterns,
- inventory controls that prevent shrink and write-offs.

These controls rarely get applause. They also quietly prevent the kind of fraud or error that forces a restatement, triggers lender concern, or derails a fundraising story. Protection is often just good operating hygiene with a CFO's insistence on consistency.

How to choose what to protect first: focus on tail risk and plan integrity

Protection budgets can balloon if you let every function chase their worst nightmare. A CFO needs a prioritization lens that is both financially rigorous and emotionally intelligent (because people's fears are often rational, just not ranked).

A practical decision filter:

- *Tail risk:* Could one event create a loss bigger than the company can absorb without emergency financing?
- *Plan integrity:* Could normal volatility (FX, rates, key input costs) repeatedly distort results enough to break trust with stakeholders?
- *Control leverage:* Is there a relatively cheap action that materially reduces probability or impact?

Tail risk gets insured and engineered around. Plan integrity gets hedged and contractually managed. High-leverage controls get implemented fast.

A short case: the quarter that got stolen by ``unmanaged" volatility

A mid-sized SaaS company begins selling into the UK and invoices in GBP because buyers prefer it. Sales celebrates: close rates rise. Finance celebrates: bookings look great. Then GBP weakens, and the next quarter's collections come in light relative to USD expectations. The company misses its revenue and margin narrative—not because customers churned, not because product failed, but because FX moved.

The company reacts emotionally:

- Sales demands higher prices, which slows deals.
- Finance tightens spend, which slows product delivery.
- Leadership spends the earnings call explaining currency, which nobody enjoys.

A simple forward hedge program—limited to high-confidence exposures—would have turned that volatility into a known cost. The company would still face competitive realities, but it would stop letting the FX market grade its operating performance.

That's the CFO's armor at its best: not dramatic heroism, just removing randomness from the scorecard.

Actionable takeaway: build a one-page Protection Stack and fund it like a portfolio

Create a one-page "Protection Stack" that sits next to your cash forecast and operating plan. Keep it brutally simple and updated quarterly.

Then do three decisions immediately:

- Pick *one* volatility exposure (FX, rates, commodities) that has been distorting results and implement a basic hedge policy within 30 days.
- Review *one* insurance policy for definitions and exclusions that could betray you on a claim, and fix it before renewal.
- Identify the *single biggest operational single point of failure* and fund a mitigation that reduces downtime risk measurably this quarter.

Armor works when it's worn before the arrow hits. The CFO's advantage is simple: you can see the balance sheet behind the bravado—and you can buy the kind of certainty that lets the business take the risks it actually wants.

7.3 Assurance and Compliance: Verification and Strategy

The email subject line is always polite. The feeling it produces is not.

"Request for PBC items."

"Notice of examination."

"We have a few follow-up questions."

Assurance and compliance are the parts of finance where your company gets graded by strangers. Not on potential. Not on effort. On evidence. If your systems are clean, this feels like a routine physical. If they're not, it feels like you're being audited by a microscope.

Here's the twist: the best operators don't treat auditors and tax authorities as enemies. They treat them as a forcing function. The process that makes you audit-ready tends to make you *company-ready*: tighter data, clearer ownership, fewer "spreadsheet miracles," and decisions that can survive scrutiny. Compliance becomes strategy when it reduces uncertainty, preserves options, and builds stakeholder trust at the exact moments you most need it.

Assurance is not "someone checking your math." It's someone checking your system.

Most non-finance leaders think an audit is about whether the financial statements add up. That's the visible part. The deeper part is whether your process for producing the numbers is reliable.

That's why you'll hear the term *ICFR*: Internal Control over Financial Reporting. In plain language, ICFR is the set of people, processes, and controls that make your financial reporting dependable. It's how you prevent (or detect in time) errors, manipulation, and omissions that could materially mislead someone relying on your statements.

There's a humbling concept at the center of assurance: *reasonable assurance*. Not perfect assurance. Not "no errors ever." Reasonable means the system is designed and operated well enough to reduce the risk of material misstatement to an acceptably low level. That framing matters because it stops the two worst behaviors:

- building a control program so heavy it strangles the business, and
- building one so casual it collapses the first time a key person takes vacation.

Auditors aren't asking you to be flawless. They're asking you to be *credible*.

Internal Audit vs. External Audit: your critical friend and the referee

Internal Audit and External Audit sound similar until you've lived with both.

Internal Audit is inside the company. When it's healthy, it acts like a critical friend: it looks for weak signals, tests whether controls actually work, and tells you uncomfortable truths early-while you still have time to fix them quietly. Internal Audit can also be a strategic partner when you're scaling: it helps you build controls that fit the business rather than controls pasted from a much larger company.

External Audit is the referee. External auditors are independent, and their job is to provide assurance to outsiders-investors, lenders, regulators-that the financial statements are fairly presented under the relevant accounting framework. They don't exist to make your life easy. They exist to protect the integrity of the capital markets (and, practically, to protect their license and reputation).

If Internal Audit says, "This process is risky," External Audit says, "Prove it isn't."

Treat that difference with respect and you'll have fewer surprises.

The Audit Committee: the room where trust is manufactured

When people talk about "governance," it can sound like ceremonial paperwork. The Audit Committee is where governance becomes operational.

The committee's power is simple: it creates a credible channel between the board and the mechanics of reporting. It hires and oversees the external auditor. It expects management to take controls seriously. It asks the question management is sometimes too busy-or too optimistic-to ask: "How do we know?"

A strong relationship with the Audit Committee is not about constant updates. It's about clarity and candor:

- What are the biggest reporting risks right now?
- Where are you relying on heroics?
- What changed in the business that could break last year's controls?
- What would constitute a "bad surprise," and what are you doing to prevent it?

The CFO who can answer those questions with specifics-owners, timelines, evidence-builds a kind of institutional confidence that pays off later. When you need to make a hard call (a write-down, a restatement risk decision, an acquisition accounting position), trust is the currency that buys you time and support.

Material weakness: the phrase you never want to learn under pressure

There's a reason "material weakness" makes executives sit up straighter than "variance" or "process gap." A material weakness in ICFR means there is a reasonable possibility that a material misstatement of the financial statements will not be prevented or detected on a timely basis. That's not about whether the current statements happen to be wrong; it's about whether your system is strong enough to be relied upon.

This distinction matters because it changes how you manage remediation. You can't "explain away" a material weakness with "but the numbers look fine." The point is that you got lucky, and luck is not a control.

A CFO who treats remediation as a project with deadlines, testing, and ownership-rather than a vague intention-protects the company's credibility. Credibility is expensive to rebuild and cheap to preserve.

Audit readiness: reduce fees, reduce drama, reduce the temptation to hide

Audit costs feel like taxes: unavoidable, mildly annoying, and somehow always higher than you wanted. But audit fees often reflect the condition of your underlying data and discipline. When the close is chaotic, auditors respond rationally: more testing, more questions, more time. When the documentation is thin, they don't "trust you more." They expand scope.

Audit readiness is therefore not just compliance. It's operating efficiency.

The CFO move is to stop thinking of audit readiness as an annual event. Treat it like you treat cash: monitored, forecasted, and managed continuously.

A practical audit-ready company has:

- a close calendar that is real (not aspirational),
- reconciliations that are performed and reviewed on time,
- documented policies for judgment areas (revenue recognition, capitalization, reserves),
- a clean trail from source systems to reported numbers,
- and evidence that controls operated, not merely existed.

The "evidence" part is where teams stumble. People do the work but don't capture proof in a way that survives scrutiny. An auditor can't audit your memory. Evidence is your shield.

The PBC list is not a scavenger hunt; it's a map of what your auditors fear

"PBC" (Provided By Client) requests can feel random until you see the pattern: auditors ask for what reduces their uncertainty. The requests are essentially a heat map of risk in your reporting.

If you treat PBC as a last-minute fire drill, you train auditors to believe your process is brittle. If you treat it as a structured deliverable process-with clear owners and standardized support-you shrink both the timeline and the tension.

A CFO-grade tactic: build a PBC library. Not a folder full of last year's PDFs, but a repeatable package:

- templates for rollforwards (debt, equity, fixed assets),

- standard tie-outs (trial balance to financial statements, subledger to GL),
- memos for key judgments (why the reserve is adequate, why revenue is recognized as it is),
- and a version-controlled way to deliver documents.

You're not just making auditors happy. You're building institutional memory so the company isn't held hostage by whoever happened to build the spreadsheet three years ago.

The fastest way to lose time: arguing about scope instead of improving the system

Many teams try to "manage the audit" by negotiating requests down. Sometimes that's appropriate-auditors can be inefficient. But a consistent pattern of defensiveness has a cost: auditors interpret it as risk. Risk triggers more work. More work costs money. The cycle feeds itself.

A better strategy is to attack root causes:

- If auditors are re-testing basic reconciliations, your controls are not designed or evidenced in a way they can rely on.
- If auditors are drilling into revenue contracts one by one, your contract review and revenue memo process is likely inconsistent.
- If auditors distrust system reports, your IT controls and data lineage may be weak.

Spend energy on system credibility, not debate. Debate is theater; credibility is leverage.

Tax strategy: the line between smart and stupid is documentation

Tax has a branding problem. People hear "tax strategy" and imagine aggressive schemes, risky structures, and headlines. In practice, most CFO-level tax strategy is painfully wholesome: don't overpay, don't underpay, don't get surprised, and don't make future transactions harder because today's tax position was sloppy.

The most expensive tax is what you might call the *stupidity tax*: missing credits and incentives you legitimately qualify for, misclassifying something obvious, or failing to document positions so thoroughly that you lose them in an examination.

Three categories where smart tax management routinely creates real value:

- *Incentives and credits:* R&D credits, investment incentives, location-based programs, energy or manufacturing credits depending on jurisdiction and activity.
- *Entity and transaction design:* how you structure contracts, intercompany charges, and ownership so you aren't paying avoidable friction later.
- *Operational controls:* sales tax/VAT compliance, payroll tax accuracy, information reporting-areas where small errors can scale into penalties.

Tax becomes strategic when it's integrated into how the business operates, not bolted on after contracts are signed and invoices are issued.

The Tax Control Framework: treat tax like a process, not a series of heroic saves

A Tax Control Framework (TCF) is exactly what it sounds like: a system of governance, controls, and documentation that makes your tax outcomes repeatable and defensible.

Why it matters: tax authorities care less about your internal intentions and more about whether your processes produce reliable results. A TCF makes your posture legible. It can also reduce the frequency and intensity of disputes, because you can show your work like a disciplined adult.

A CFO doesn't need a massive TCF to get value. Start with the core:

- *Governance:* who owns tax decisions, who reviews, who approves, and how issues are escalated.
- *Controls:* what prevents common errors (nexus determinations, exemption certificate handling, intercompany agreements, permanent vs. temporary differences).
- *Data:* where the numbers come from and how they reconcile to the books.
- *Documentation:* memos for significant positions, maintained contemporaneously.

The payoff is not just reduced risk. It's speed. Deals close faster when diligence questions have answers. Audits run smoother when support is coherent. Leadership can make decisions without feeling like tax is a black box.

Materiality: the audit concept that doesn't copy-paste cleanly into tax

Finance teams often try to use financial statement materiality as a proxy for tax risk. That's tempting-and frequently wrong.

Financial materiality asks: "Would this change a reasonable investor's view of the financial statements?" Tax authorities ask: "Was the correct tax paid under the rules?" They care about compliance at the transaction and jurisdiction level, and thresholds can be different. A position that is immaterial to consolidated financial statements can still be meaningful in a specific state, country, or tax type-especially when penalties and interest compound.

A CFO should therefore think in two materialities:

- *Financial statement materiality:* what could distort reported results.
- *Tax authority materiality:* what could trigger disputes, penalties, or reputational damage in a jurisdiction.

Treat them as related but not identical, and your risk assessments stop making accidental blind spots.

Managing tax examinations: be cooperative without being casual

When a tax authority comes calling, the first instinct in many companies is either panic or bravado. Neither helps. The right stance is professional, prepared, and precise.

Prepared means:

- you can reconcile tax returns to the books cleanly,
- you have support for significant positions (not recreated in a weekend),
- and you know where judgment was applied.

Precise means:

- you answer the question asked, not three extra questions you accidentally volunteer,
- you maintain a clear record of requests and responses,
- and you keep internal alignment so the company tells one coherent story.

Cooperative does not mean disorganized. A messy response process invites more scrutiny because it signals that the system behind the numbers may be messy too.

Where assurance and tax connect: a single source of truth beats cleverness

The same weaknesses that drive ugly audits drive ugly tax outcomes:

- inconsistent master data (entities, customers, products),
- unclear revenue policies,
- weak documentation of judgments,
- and reconciliation gaps between subledgers and the GL.

Fixing those issues is operational excellence disguised as compliance. It also makes you faster. When reporting becomes reliable, leadership stops burning cycles on re-litigating the numbers and starts using them.

There's also a human side: strong assurance and tax processes reduce the organization's reliance on heroes. Heroes burn out. Systems scale.

A CFO's playbook for a low-drama Audit Committee relationship

If the Audit Committee only hears from you when something is on fire, you've trained them to assume there are always hidden fires. That makes meetings tense and decisions slow.

A healthier operating model:

- Bring forward the top reporting risks early, in plain language.
- Show remediation plans with dates and owners.
- Be candid about judgment calls and why you landed where you did.
- Don't drown them in detail; offer depth on demand.

You are not selling. You are informing. The committee's job is oversight; your job is to make oversight possible without turning the company into a bureaucracy.

Actionable takeaway: implement an ``Evidence First" system in 45 days

If you want compliance to feel less like a burden and more like an asset, build one habit: *evidence first*. The habit is simple: every important number and position should have support that is created as part of the process, not reconstructed later.

A 45-day implementation plan that changes behavior quickly:

1. Create a *Top 12 Evidence List*: the reconciliations, memos, and rollforwards that drive your financial statements and major tax positions (revenue, AR, reserves, equity, debt, payroll taxes, sales tax/VAT, intercompany).
2. Assign an owner to each item and define what "good evidence" looks like (tie-outs, approvals, version control).
3. Standardize templates for judgment areas (revenue recognition conclusions, reserve methodology, uncertain tax positions, capitalization policies).
4. Build a shared, version-controlled repository with a naming convention that makes sense to someone new.
5. Run a mock request drill: pick five items from your likely PBC/-tax audit requests and see how fast you can produce clean support. Fix what breaks.

The result is not just smoother audits and fewer tax surprises. It's a company that can prove what it knows. In a world where trust is fragile and scrutiny is constant, the ability to prove it is a competitive advantage.

Chapter 8

Building the Engine: Systems, Data, and Team

You can have the perfect strategy, but if your engine-your people, data, and systems-is held together with duct tape, you won't get far. This chapter moves beyond org charts and software demos to show you how to build a finance function that runs on autopilot, turns chaos into clarity, and serves as the ethical backbone of the entire company.

8.1 The Finance Operating Model

The telltale sign of a broken finance operating model is a heroic CFO. Every question—*Can we hire? Can we discount? Can we renew that vendor? Why did margin dip?*—funnels to one person because the system can't answer it, the process can't route it, or the team doesn't have the decision rights to act. The calendar fills with "quick checks" that are never quick, approvals that are really audits, and meetings that exist mainly to reconcile whose spreadsheet is "the real number." If the company grows, the CFO's day doesn't just get busier; it becomes structurally impossible.

A finance operating model is the architecture that prevents that outcome. It's less about boxes on an org chart and more about the choreography of work: where transactions land, who owns a number, how exceptions are handled, and which decisions are made once (in policy) instead of made repeatedly (in email). Get it right and finance feels like power steering—light touch, high control. Get it wrong and finance becomes a toll booth: everything stops, nothing flows, and the organization learns to drive around you.

The tricky part is that "better" isn't synonymous with "more complex." Many teams build accidental complexity: a new tool to patch a reporting gap, a new approval to prevent a mistake, a new spreadsheet to bridge systems that don't talk. Each patch feels rational in the moment. Over time, those patches become your operating model: brittle, expensive, and entirely dependent on a few people who know the secret handshakes. The goal isn't to eliminate judgment. The goal is to reserve judgment for the moments that deserve it.

Start with the work, not the org chart

The operating model begins with a blunt inventory of finance work, because different work needs different design.

Transactional work is high-volume, rules-based, and intolerant of error: vendor onboarding, invoice processing, cash application, payroll journal entries, expense reports, billing, revenue recognition routines, and reconciliations. The win condition here is boring excellence-fast, consistent, well-controlled.

Control work keeps the machine honest: close management, accounting policy, internal controls, segregation of duties, audit support, and compliance. The win condition is confidence: management, auditors, and the board believe the numbers without a long speech.

Decision support is where finance earns its seat at the table: forecasting, unit economics, pricing, capital planning, customer profitability, and investment cases. The win condition is influence: the business changes behavior because finance made the trade-offs legible.

Too many teams try to run all three categories with the same people, the same tools, and the same cadence. That's how you end up with FP&A analysts approving invoices, controllers building pricing models, and the CFO personally reconciling a bank account before a board meeting. Not because anyone is incompetent-because the model forces it.

A practical way to begin is to map your recurring work against two axes: *volume* and *judgment*. High-volume/low-judgment work wants standardization, automation, and clear service levels. Low-volume/high-judgment work wants proximity to decision-makers, strong business context, and time to think. The operating model is the set of choices that puts each kind of work in the environment where it can succeed.

Three ways to place the work: embedded, shared, or expert

Most finance organizations end up using a blend of three "work placement" patterns. The mistake is treating these as trendy labels. They're simply answers to: *Where should this work live so it moves fast and stays controlled?*

Embedded finance means finance partners sit close to the functions they support-Sales, Product, Marketing, Operations. This is where decision support thrives because context is oxygen. The embedded model is also where finance can shape behavior early: pricing before the quote goes out, hiring plans before offers are written, contract terms before legal redlines become a week-long sport. Embedded finance is not "being nice to the business." It's moving influence upstream, where choices are still cheap to change.

Shared services consolidates transactional processing in one team (sometimes centralized, sometimes distributed but standardized). The economics are obvious-repeatable work improves with repetition-but the strategic benefit is often bigger: shared services creates a single way of doing things. When a company says, "We do expense reports *like this*," it stops renegotiating the process every time a new leader arrives with a new preference.

Centers of Excellence (*CoEs*) are small expert teams that own specialized knowledge and standards: revenue recognition, technical accounting, tax, treasury, pricing, data/analytics, or systems. CoEs prevent the "local workaround" problem. Without them, every business unit invents its own version of revenue timing, customer definitions, or KPI logic. With them, the business moves faster because it's not arguing about definitions in every meeting.

The blend depends on the business. A global manufacturer might centralize payables and keep plant controllers embedded. A SaaS company might have a small shared services core, a revenue CoE, and strong embedded FP&A partners for go-to-market. The principle is stable: standardize what should be standard, embed what must be contextual, and centralize expertise that should not be reinvented.

Decision rights: the difference between ``empowered" and ``chaotic"

You can hire brilliant people and still create bottlenecks if decision rights are fuzzy. Fuzzy decision rights create two predictable behaviors: either everyone escalates ("I don't want to be wrong"), or everyone improvises ("I didn't know I needed permission"). Both are expensive.

Decision rights sound abstract until you write them down. Here's what "writing them down" looks like in finance language:

- *Who can commit the company to spend?* (Not "who can request." Who can obligate.)
- *Who can change a forecast?* (And what constitutes "the forecast" versus a scenario.)
- *Who owns customer terms?* (Payment terms, billing cadence, refunds, credits.)
- *Who defines KPIs?* (Gross margin, ARR, churn, contribution margin-definitions, not just dashboards.)

- *Who can approve exceptions?* (And what's an exception worth escalating.)

Good operating models treat decision rights as a product: designed intentionally, tested under pressure, and improved over time. The CFO's role isn't to be the decision. It's to design the system that produces decisions with the right speed and the right control.

A useful mental model: move decisions from *people* to *policies* where you can. If every non-standard discount requires CFO review, you don't have a discount policy-you have a CFO-shaped policy. A policy can include guardrails (floor price, approval tiers, required ROI logic) and still preserve flexibility. The point is to make the routine routable.

Process architecture: stop building ``approval mazes"

Organizations often react to a control failure by adding an approval. An expense was mis-coded? Add an approval. A vendor invoice was fraudulent? Add an approval. A contract had a bad clause? Add an approval. The approval count rises like plaque in an artery: slowly, then suddenly. The company still fails, just more slowly-and now everyone is annoyed.

Process architecture is the discipline of designing workflows that handle routine work automatically and route the *right* exceptions to the *right* humans. The architectural question isn't "How do we prevent mistakes?" It's "How do we create a system where mistakes are hard to make, easy to detect, and quick to fix?"

A few design patterns show up in every high-functioning finance team:

1) *Separate initiation from approval.* The person who wants the thing shouldn't be the person who approves the thing. This isn't cynicism; it's basic control hygiene. It also protects relationships. When managers approve spend, finance isn't forced into the role of "bad cop" on every purchase.

2) *Treat coding as data entry, not a debate.* The chart of accounts should be usable by normal humans. If an employee needs a mini-course to pick the right department and project code, you haven't created discipline-you've created error. A well-designed COA and clear default rules reduce rework dramatically.

3) *Design for throughput at the constraint.* The constraint is rarely the AP clerk. It's usually vendor onboarding, PO creation, receipt confirmation, or contract review. If invoices pile up, the "fix" is often upstream: purchase orders that don't exist, mismatched receipts, unclear approvers. Finance becomes fast when the business's prerequisites are fast.

4) *Build an exception taxonomy.* Not every exception deserves escalation. Define what's truly risky (new vendor + unusual bank change + rush payment) versus merely annoying (missing PO number on a low-dollar recurring bill). When exceptions are categorized, you can automate the low-risk ones and focus human attention where it matters.

This is where operating model choices become cultural. A clean process tells the organization, "We respect your time." A messy one tells them, "We'll take your time whenever we're nervous."

The tech stack as a layer cake (and why the frosting matters)

Finance technology is often bought the way people shop hungry: impulsively, with regret later. A sales demo shows a dashboard that answers a question you've been dodging for months, and suddenly you're signing something that "integrates with everything." Six months later, you've got another login, another data pipeline, and another reconciliation-the opposite of what you wanted.

The best mental model for finance systems is a layer cake. The foundation is where transactions are recorded. The middle is where you control, plan, and move cash. The top is where you translate data into decisions. If you invert the cake-buying analytics before the foundation is stable-you get beautiful charts made from unreliable numbers. That's not insight; it's decor.

Place this diagram near your system selection conversations, because it forces a question that cuts through feature lists: *What layer are we buying, and what do we expect it to do for the layers above it?*

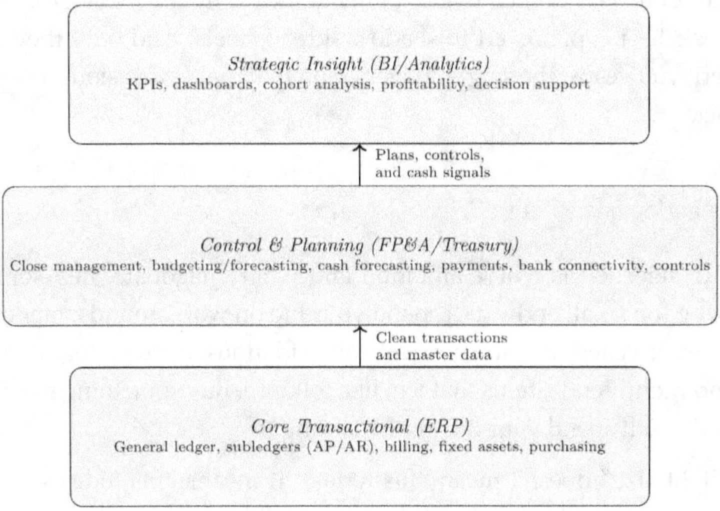

Rule of thumb: stabilize the foundation before you decorate the top.

The bottom layer-your ERP (Enterprise Resource Planning system)-is the system of record. "System of record" is a fancy way to say: if there's an argument, this is where the official answer lives. The ERP's job is not to be clever. It's to be consistent. It needs to handle your core transactions (AP, AR, GL) with clean audit trails and a chart of accounts that can survive growth.

The middle layer is where finance becomes operationally powerful. FP&A tools turn raw transactions into a living plan. Treasury tools turn bank data and cash movements into visibility, controls, and forecasting. This layer is where many CFOs first feel relief, because it reduces the most exhausting kind of work: rekeying, reconciling, and re-explaining. When planning and cash systems are designed well, finance stops being surprised by results that were already inevitable.

The top layer-BI and analytics-is where the company experiences finance as insight rather than enforcement. But BI can only be as trustworthy as the layers beneath it. A dashboard is a promise. If the promise is broken twice ("Why does this number change every week?"), people revert to shadow spreadsheets. And once they revert, you've lost the single conversation that makes decision-making fast.

Selecting an ERP: the ``right size" trap

ERP selection is where ambition and reality negotiate. Buy something too small and you'll spend your life on workarounds: manual revenue schedules, spreadsheet consolidations, brittle integrations, and month-end rituals that feel like folklore. Buy something too big and you'll spend your life implementing it.

"Right size" doesn't mean "fits today." It means "fits today's complexity with a clean path to tomorrow's." A few practical lenses keep the decision grounded:

Complexity, not headcount. A 50-person company with subscriptions, usage billing, multi-entity structure, and foreign tax exposure can be "bigger" in finance complexity than a 500-person domestic services firm. Pick systems based on complexity: revenue model, legal entity count, currencies, compliance requirements, and integration needs.

Close velocity as a design constraint. If you want a five-day close, the ERP must support it: strong subledger controls, consistent postings, disciplined master data, and reliable integrations. The system doesn't magically create speed, but the wrong system guarantees slowness.

Auditability as a feature, not an afterthought. When a business grows, the first time an auditor asks "show me the trail from invoice to financial statement," you find out whether your processes are real or improvised. ERPs differ in how well they preserve traceability across modules and how cleanly they handle permissions and approvals.

Integration realism. "Integrates with everything" often means "can be integrated with a lot of effort." The right question is: what are your *non-negotiable* integrations-billing, payroll, CRM, bank feeds, expense management-and how mature are the connectors in the wild? Integration debt is finance debt: it shows up as reconciliations, timing differences, and lost trust.

A CFO who has lived through a painful implementation learns a sobering truth: the software is rarely the main problem. The main problem is unclear process ownership. An ERP forces decisions about how the company will operate-approval flows, spend categories, customer hierarchies, revenue policies. If the organization can't make those decisions, no system can rescue you.

Layering planning and treasury tools: don't automate confusion

After the ERP, the most common buying pattern is "We need forecasting" or "We need better cash visibility." Both are legitimate needs, but they can be solved badly.

Planning tools amplify whatever planning discipline you already have. If your forecast is a negotiated story told in meetings, a planning tool will make that story faster to produce-but not more accurate. The right sequence is: define the drivers (what truly moves revenue, margin, headcount, cash), define the cadence (weekly pipeline, monthly forecast, quarterly plan refresh), and define accountability (who owns each driver). Then pick tooling that makes those mechanics easy.

Treasury tooling is similar. Cash forecasting isn't just a spreadsheet problem; it's an information problem. Cash is affected by AR collections behavior, AP payment terms, payroll cycles, tax remittances, and debt covenants. Treasury tools that connect to banks can give you near-real-time balances, but the forecast quality still depends on the operating model: who updates expected inflows/outflows, how often, and with what evidence.

The hidden win of good planning and treasury layers is not prettier reports. It's reduced *latency*. Latency is the time between reality changing and the organization noticing. A strong operating model shrinks that time so the business can steer earlier, when options are still plentiful.

Designing workflows that don't need you

A CFO bottleneck is usually created by one of three failure modes:

1) *Unclear thresholds.* People escalate because they don't know what "big" is. Fix it by defining thresholds that match risk: dollar amounts, contract length, non-standard terms, unusual vendors, margin impact. Then publish them. Then enforce them consistently.

2) *Low trust in data.* People escalate because they don't trust the number. Fix it by improving data integrity in the ERP layer, tightening definitions, and creating a repeatable close. When the numbers are dependable, fewer conversations require a referee.

3) *Too many bespoke paths.* People escalate because every case is "special." Fix it by standardizing the 80% cases into a clear path, and designing exception handling for the remaining 20%. Exceptions should be visible and measurable, not hidden in inboxes.

Think about finance workflows the way a great restaurant thinks about the kitchen. The head chef doesn't personally plate every dish. The kitchen works because recipes are defined, stations are staffed, and quality checks happen in the flow. The chef steps in for the dishes that require artistry-or when something is going wrong. Finance should feel the same: standardized where it should be standardized, and deliberately human where judgment is valuable.

Service levels: the quiet weapon of a respected finance team

Finance earns political capital when it behaves like a service organization without becoming a doormat. The tool for that is service levels: clear commitments about timing, inputs, and outputs.

A service level is not a bureaucratic artifact. It's a promise that reduces noise. "Vendor onboarding in two business days if the W-9 and bank letter are complete." "PO approvals within 24 hours for budgeted spend." "Sales commission statements by the third business day." When finance makes reliable promises, the business stops interrupting to check status. Interruptions are where bottlenecks breed.

Service levels also expose upstream dysfunction. If AP can't pay on time because the business doesn't confirm receipt, the SLA will fail—and the failure will have a name. That's good. An operating model is healthy when problems become visible quickly, not when they stay hidden behind heroic effort.

The CFO's desk as a design constraint

A simple design test: if you vanished for two weeks, what breaks?

If close breaks, you have key-person risk in accounting operations. If cash breaks, you have key-person risk in treasury and payment controls. If the forecast breaks, you have key-person risk in planning discipline. If everything breaks, you don't have an operating model-you have a set of favors you personally do for the company.

This isn't about replacing the CFO. It's about making the CFO's time scarce in the way it should be scarce: saved for decisions that change the company's trajectory, not for pushing paper through a system that lacks structure.

Actionable takeaway: run the ``bottleneck audit" this week

Block 60 minutes and do a bottleneck audit with brutal honesty. On a single page, write:

- The last 10 things that *needed* CFO involvement (not the things you chose to be involved in).
- For each, classify the root cause: (*a*) *missing policy/threshold,* (*b*) *unclear decision rights,* (*c*) *untrusted data,* (*d*) *process exception,* (*e*) *tooling gap.*
- Pick one recurring item and remove yourself from it by design: write a threshold, assign an owner, define the approval path, and set a service level.

If you do only one thing, do this: convert one repeating CFO interruption into a documented rule with an owner and a measurable turnaround time. That's how operating models are built—not with a reorg announcement, but by steadily turning heroics into architecture.

8.2 Data and Modernization

A finance team can survive almost anything except a contested number.

The meeting starts normally: revenue is up, churn is down, cash is... complicated. Then someone asks an innocent question-"Which figure are we using?"-and the room splits into tribes. Sales has a spreadsheet. FP&A has a model. Accounting has the ERP report. The BI dashboard has a fourth answer that looks confident enough to be wrong. Ten minutes later, nobody is talking about customers or strategy. They"re arguing about whose data is "real," and the CFO is cast as judge, jury, and reluctant historian.

That is *Excel Hell*: not Excel itself, but the social reality of unmanaged data. It"s the tax you pay when information is allowed to behave like gossip—passed around, edited, embellished, and re-shared until it becomes impossible to prove what"s true. Modernization is the act of turning gossip back into records.

The Single Source of Truth isn't a database. It's a contract.

"Single Source of Truth" gets sold like a piece of software, as if you can buy a platform and wake up the next morning to harmony. In practice, it"s a contract the company makes with itself:

- *These are the official definitions.*
- *These are the official systems where those facts live.*
- *These are the people who have the right-and the responsibility-to decide when a definition changes.*

Notice what"s missing: it"s not a promise that there will only be one system. Companies will always have multiple systems because the world is messy: CRM for pipeline, ERP for accounting, payroll for wages, billing for invoices, bank portals for cash, support tools for tickets, and so on. The point isn"t to eliminate plurality. The point is to eliminate *ambiguity*.

A good Single Source of Truth feels boring because it turns drama into routine. When someone asks, "What was ARR last month?" nobody replies, "Define ARR." The company has already defined ARR. The question becomes: "Do you want it gross or net of credits?"-a sophisticated question, not a foundational one.

The CFO's modernization job: reduce reconciliation as a way of life

Reconciliation is a healthy activity when it"s a safety check. It becomes a disease when it"s the primary method of producing truth.

If your monthly rhythm includes a heroic "recon week" where people manually tie out sales to billing to cash to revenue to deferred revenue, that"s not a rite of passage. That"s a signal that the organization is running on mismatched clocks. Each system is telling time differently, and finance is paid (in stress) to translate.

Modernization doesn"t mean "less work." It means *less translation work*. Translation is where errors hide: copy-paste, filtering, rekeying, and the subtle moment when an analyst decides a transaction "probably belongs" in a category because there isn"t a clean rule. Every manual bridge becomes a control risk and a morale problem. Your best people weren"t hired to be human middleware.

Data governance: the unglamorous superpower

Data governance sounds like a committee that meets to discuss naming conventions. Done badly, that"s exactly what it is. Done well, it is a set of decision rights that protects speed.

Governance answers three questions that every growing organization eventually asks, whether it admits it or not:

1. *Who gets to define the meaning of a number?*
2. *Who is accountable when that number is wrong?*
3. *How does a definition change without breaking everything downstream?*

The most common governance failure is pretending that "data" belongs to IT. IT can host it, secure it, and move it around. But ownership-the authority to decide what a customer is, what a booking is, what counts as "active," when revenue is recognized-must sit with the business. Finance often ends up owning or co-owning critical definitions because finance is where numbers become commitments: to investors, lenders, boards, and payroll.

A useful phrase here is *decision rights*. Governance is not a poster; it"s a map of who decides what. When governance is missing, two things happen. First, definitions drift. Second, every disagreement becomes a bespoke negotiation. The company slows down because truth is negotiated in real time.

Critical Data Elements: pick the few that can ruin your month

Trying to govern all data is like trying to mop the ocean. The move that makes governance practical is focusing on *Critical Data Elements* (CDEs): the small set of fields that, if wrong, create real damage.

CDEs differ by business model, but the pattern is consistent. The critical fields are the ones that tie money to accountability:

- Customer identifiers and hierarchies (who is paying, and who is actually using)
- Product/SKU or service package definitions (what was sold)
- Contract start/end dates and billing cadence (when money should move)
- Price, discount, and approval references (why margin looks the way it looks)
- Tax and entity fields (where obligations land)
- Vendor master and bank details (where money goes)
- Employee cost centers and job codes (how labor is allocated)

If you get these right, a huge portion of downstream reporting becomes trustworthy *by construction*. If you get them wrong, you can build the prettiest dashboard in the world and it will still be a high-resolution hallucination.

The CFO doesn"t need to personally manage these fields. The CFO needs to insist that someone does-and that the role comes with authority. "Accountable" without authority is just blame waiting to happen.

Data quality: accuracy is only one of the villains

Data quality is often treated as a synonym for "accuracy." Accuracy matters, but it"s not the only way data can betray you. In finance operations, the most damaging issues are often:

Completeness: fields that should be required are left blank, so transactions fall into "miscellaneous" buckets that never get cleaned up.

Consistency: the same concept is represented differently across systems (NetSuite "customer" vs. Salesforce "account," product families that don"t match, regions that mean different things in different tools).

Timeliness (and currency): the information is correct, but it"s stale. A payment status updated three days late turns cash forecasting into superstition. A headcount report that lags hires by two weeks makes "run-rate" calculations look precise while being useless.

Validity: values exist but violate rules (end date before start date, negative quantities, discounts beyond policy).

The point of naming these dimensions isn"t to build a glossary. It"s to make problems diagnosable. When someone says "the data is bad," that"s an emotional statement. When they say "the data is complete but not current," that"s an operational problem with a fix.

Master data: where control actually starts

If modernization is a factory, master data is your raw material. Master data is the set of shared reference lists the company relies on: customers, vendors, products, chart of accounts, cost centers, locations, entities, and sometimes projects.

If master data is unmanaged, finance spends its life on two kinds of rework:

- *Deduplication rework*: "Acme Inc.", "ACME", "Acme Corporation" are three customers until someone manually merges them-after you"ve already produced a report that counts them separately.
- *Reclassification rework*: the wrong department, product, or region was selected at the point of entry, so the month-end close becomes a massive recoding project.

Modernization here looks less like AI and more like discipline:

- Make CDE fields required at creation (customer, vendor, product).
- Limit who can create or edit master records.
- Build approval flows for sensitive changes (bank details, payment terms, revenue treatment flags).
- Create a "golden record" concept: one approved representation that downstream systems inherit.

This can feel restrictive until you remember the alternative: a company where anyone can create a vendor and change bank details is not "agile." It"s *exposed.*

The pipeline mindset: stop treating reporting as a one-off project

A mature finance organization treats reporting as a pipeline, not a pile of requests. Pipelines have inputs, transformations, controls, and outputs. Piles have heroes.

When reporting is a pile, it looks like this:

- A board deck is created by hand each month.
- KPIs are recalculated in a spreadsheet because last month"s spreadsheet is "almost right."
- Every new executive asks for "their version" of the metrics.
- Nobody can answer "where did this number come from?" without opening five tabs and calling someone.

A pipeline looks different:

- Data sources are defined (ERP, CRM, billing, payroll, bank).
- Transformations are documented (how ARR is derived, how churn is calculated, how revenue is allocated).

- Controls exist in the flow (reconciliations, variance checks, anomaly flags).
- Outputs are consistent (dashboards, management reporting, forecasts).

The goal isn"t perfection. The goal is *repeatability*. Repeatability is what allows improvement. You can"t improve a process that only exists in someone"s head.

Automation: the right target is ``low judgment, high volume''

Automation has a way of attracting the wrong ambitions. Teams try to automate decisions before they automate data entry. They try to build predictive models while still fixing basic coding errors at month-end. The result is a Frankenstein machine: expensive, fragile, and ultimately distrusted.

A cleaner rule: automate the work that is high-volume and rules-based, and make the rules explicit. In finance, that usually means:

- Invoice capture, coding suggestions, and matching (PO, receipt, invoice)
- Expense processing (policy checks, duplicate detection, category suggestions)
- Bank reconciliations and cash application
- Journal entry creation for routine accruals
- Intercompany balancing for standard transactions

These are not glamorous tasks, but they are where time evaporates. They are also where error rates compound: one miscoded invoice becomes a misclassified department spend, which becomes a misleading variance, which becomes a bad decision about budget cuts. Automating the base layer is how you protect decision-making upstream.

Smart automation is not ``no humans.'' It''s ``humans in the right moments.''

The most practical modern automation pattern looks like this:

- *Capture*: documents and data come in (emails, PDFs, portal uploads, receipts).
- *Extract and classify*: the system reads the content (vendor, amount, date, category) and proposes what it is.
- *Match and validate*: it checks against rules and existing records (is vendor approved, does it match a PO, is the amount within tolerance, does it violate policy).
- *Route exceptions*: ambiguous, high-risk, or non-compliant items go to humans with context and a recommended action.
- *Learn*: recurring exceptions become new rules, better mappings, better training data.

This is the heart of "smart" automation: not a fantasy of removing humans, but a disciplined design for when humans are needed. Most finance operations don"t fail because humans exist. They fail because humans are asked to do what machines do well (copying, checking, remembering rules), leaving no time for what humans do well (judgment, negotiation, design).

When automation is designed correctly, it changes the emotional experience of the job. AP stops being a race against inboxes. Expense processing stops being a weekly ritual of petty policing. The team gets to spend its attention on exceptions that matter: new vendors, unusual terms, fraud signals, or spend that contradicts strategy.

Controls don't disappear when you automate. They move.

Automation can reduce risk, but it can also move risk into places that feel invisible.

Manual processes have obvious failure points: a missed approval, an incorrectly coded invoice, a late reconciliation. Automated processes have different failure points: a rule that"s wrong, a mapping that silently misclassifies, a permissions setup that allows the wrong person to approve, an integration that fails and backfills bad data.

The control mindset that keeps modernization safe is simple: *trust, but verify*. Build checks that run automatically and loudly:

- Duplicate payment detection
- Vendor bank detail change alerts
- Spend anomalies by vendor/category/department
- Reconciliation breaks between subledgers and the GL
- Forecast vs. actual variance thresholds that trigger investigation

And keep a human review layer for the decisions that could hurt you: large payments, new vendors, unusual contract terms, and anything that affects revenue recognition or cash covenants.

Modernization isn"t a trade of speed for control. It"s a chance to get both—if you design controls as part of the system rather than as an after-the-fact apology.

A modernization sequence that actually works

Modernization projects fail when they start with tools instead of truths. A sequence that tends to work in the real world is:

1. *Standardize definitions*: pick the official metrics and write the definitions in plain language.
2. *Assign data owners*: for each CDE, name the accountable business owner and the operating steward.
3. *Clean the master data*: customers, vendors, products, COA, cost centers-fix the raw material.

4. *Stabilize integrations*: ensure systems sync reliably, with monitoring and reconciliation checks.
5. *Automate the volume*: AP/expenses/reconciliations-free time and reduce error.
6. *Scale insight*: dashboards and analytics become credible because the foundation is credible.

This is less exciting than "implement AI." It"s also far more likely to produce a finance organization that can operate at speed without losing its grip on reality.

The cultural shift: from ``spreadsheet craftsmanship" to ``system design"

There"s a human side to Excel Hell that rarely gets acknowledged: people become proud of their spreadsheets. They should be. Those spreadsheets are often feats of ingenuity built under pressure, with incomplete systems and unreasonable deadlines. Taking them away can feel like taking away someone"s proof of competence.

The respectful modernization move is to redefine what excellence looks like. Excellence is no longer "the person who can fix the model at 1 a.m." Excellence becomes "the person who designs the process so the model doesn"t break," or "the person who improves data quality so the analysis is trusted," or "the person who turns an exception pattern into a durable rule."

Finance leaders who get this right make modernization feel like a promotion, not a replacement. The best spreadsheet builders often become the best system thinkers-once you give them permission to stop being firefighters.

Actionable takeaway: build a one-page ``Finance Data Constitution''

Create a one-page document that can live on a wiki, get revised, and be pointed to in meetings. Title it *Finance Data Constitution*. Keep it short enough that people actually read it.

Include:

- *Five Critical Data Elements*: pick the five fields most likely to distort revenue, margin, or cash (e.g., customer ID, product, contract start date, payment terms, vendor bank details).
- *Owner + steward for each*: owner is accountable for meaning; steward runs the day-to-day hygiene.
- *Official definitions for three metrics*: choose the three most argued-about numbers (ARR, gross margin, cash burn, churn-whatever causes the most heat) and define them in plain language.
- *Quality rules*: one measurable rule per CDE (e.g., "100% of new customers must have a legal entity name + payment terms," "no vendor bank changes without dual approval," "contract start/end dates required for all subscription items").
- *Change process*: one sentence on how definitions change (who approves, where it"s documented, and how downstream users are notified).

Then do the real test: in the next meeting where someone challenges a number, point to the Constitution before you debate. If it doesn"t settle the dispute, you"ve learned exactly what to fix. That"s modernization in its most practical form: turning recurring arguments into a stable operating system.

8.3 Leadership and Ethics

The fastest way to learn what a finance team really does is to listen for the phrase "Can you just...?"

"Can you just pull a quick report?" becomes a week of reconciling definitions. "Can you just approve this invoice?" becomes a quiet debate about whether the vendor is real. "Can you just tell me if we can afford it?" becomes a referendum on whose forecast is credible. Over time, the company trains finance to be the adult in the room for every messy decision—and then acts surprised when finance is stretched thin and perpetually late.

Leadership in finance is the art of breaking that pattern without breaking trust. You want a team that is helpful, but not enabling; responsive, but not reactive; flexible, but not porous. The ethical dimension is inseparable from the leadership dimension because finance is where reality gets written down. When the numbers are massaged, delayed, or selectively interpreted, the business isn't merely confused—it's lied to. And even "small" lies compound quickly because the systems you built to create speed will happily scale dishonesty, too.

The modern finance team isn't one team. It's three gears that must mesh.

High-performing finance functions don't look like a ladder of seniority. They look like a machine with distinct gears that turn at different speeds:

The reliability gear keeps the company grounded: accounting operations, close management, core controls, and technical accounting. This gear values precision, repeatability, and professional skepticism—the healthy habit of asking "How do we know?" before concluding "We know."

The liquidity gear makes cash visible and intentional: treasury, working capital, payments governance, and cash forecasting. This gear values timing, counterparty risk awareness, and the ability to spot trouble early, when it is still a decision instead of an emergency.

The decision gear turns data into choices: FP&A, business partnering, pricing, unit economics, and investment analysis. This gear values context, storytelling with numbers, and the courage to challenge a popular plan with an unpopular fact.

When these gears blur, finance becomes brittle. Reliability gets sacrificed for speed. Liquidity gets treated as a weekly spreadsheet ritual. Decision support becomes a slide factory. The fix isn't to create more layers of management; it's to assign clear ownership for each gear and design handoffs that feel natural rather than political.

Business partnering: the difference between ``presenting numbers" and changing outcomes

Finance business partnering is often described as "being strategic," which is vague enough to be meaningless. In practice, a finance partner earns the title when their work changes what the business does, not just what it knows.

A finance partner in Sales doesn't merely report performance; they shape it. They help decide which segments to pursue, where discounting is genuinely strategic versus simply habitual, and which deal terms create future churn disguised as present revenue. They build simple, durable economics that a sales leader can use under pressure: contribution margin after commissions, implementation cost reality, cash collection implications, and renewal risk.

A finance partner in Product doesn't just track spend; they translate product bets into financial language. They clarify whether a feature is a retention play, a pricing power play, or a cost-to-serve play—and what success actually means in measurable terms. They force clarity on trade-offs: "If we build this, what do we stop building?"

The ethical dimension shows up here, too: a finance partner's job isn't to tell Product what story investors want. It's to make sure the company is telling itself the truth about what it is building and why.

This is why the "scorekeeper" stereotype is both unfair and dangerous. Unfair because most finance teams are trying to help. Dangerous because it tempts finance leaders to over-correct into cheerleading. Finance is not the Department of No, but it also can't become the Department of Whatever. The partner stance is *co-pilot*: committed to the destination, relentless about instrument readings.

What to hire for: curiosity beats charisma

Many finance teams hire "strong communicators" and then wonder why analysis quality declines. Communication matters, but the engine is curiosity—the instinct to go one layer deeper:

- When revenue is up, curiosity asks: "Which customers? Which products? What changed in pricing or mix?"
- When churn is stable, curiosity asks: "Is that because retention improved, or because growth masked the losses?"
- When cash is tight, curiosity asks: "Is this a timing issue, a margin issue, or a working capital issue?"

Curiosity is also a control muscle. It's what prompts someone to notice that an "urgent" vendor payment request has a new bank account, or that a quarter-end revenue entry looks oddly round, or that a KPI improved in a way that contradicts operational reality. The best ethics programs in the world are weak if curiosity is absent, because fraud and manipulation thrive in environments where nobody asks boring questions.

A useful hiring lens is to separate two capabilities that often get conflated:

Technical competence is knowing what the rules are—GAAP/IFRS basics, accruals, revenue recognition logic, internal controls, and the mechanics of statements.

Commercial competence is knowing why the business makes money—how customers buy, how products are delivered, what drives costs, and what competitors do differently.

Your team needs both. The leadership task is to build a bench where people can grow across that line rather than being trapped on one side of it.

A practical structure: clarify ``who owns the number'' versus ``who explains the number''

One of the cleanest ways to reduce friction is to distinguish ownership from narrative.

Accounting should *own* financial statement numbers: what revenue is, what expenses are, what cash is, and what the balance sheet says. Ownership means responsibility for the integrity of the record.

FP&A and business partners should *explain* those numbers in a way operators can use: what drove variance, what the forward view implies, and what choices are available.

When the same person is asked to both own and explain under time pressure, two bad behaviors emerge. Either they become overly conservative ("I won't say anything until I'm certain"), and finance loses influence. Or they start improvising explanations ("it's probably mix"), and finance loses credibility.

This division also protects ethics. A clean separation makes it harder for anyone—inside or outside finance—to pressure the team into moving numbers for narrative convenience. The record stays the record; the story stays honest.

Tone at the top: a CFO's ethics are operational, not inspirational

"Tone at the top" is often treated like a values statement. In reality it's a series of daily micro-decisions that teach the organization what is safe.

When a CFO responds to bad news with curiosity rather than rage, people bring problems earlier. When a CFO rewards someone for surfacing a control weakness, people stop hiding near-misses. When a CFO refuses a quarter-end "just this once" entry that makes results look smoother, everyone learns what the line is.

Ethics in finance is not primarily about catching villains. It's about preventing rationalizations in decent people under pressure. Pressure creates stories: *"We'll true it up next month." "Everyone does this." "It's not material." "The auditors will never notice."* A CFO's job is to make those stories socially unacceptable and procedurally difficult.

The uncomfortable truth: the CFO is often the last adult with the power to stop a bad decision that still looks "reasonable" to everyone else. Sales wants to pull forward a contract signature. Product wants to capitalize costs aggressively. Operations wants to delay recognizing an expense because the budget is tight. Any one of these can be framed as an optimization. Over time, optimizations become distortions. Distortions become reputational damage.

ICFR: internal control is the price of being believed

Internal Control over Financial Reporting (ICFR) sounds like a regulatory acronym because it is one, but the concept is simple: can the company produce financial statements that are reliable, on purpose, every time?

For public companies, this is formalized under Sarbanes-Oxley (SOX). But the spirit matters even for private companies, especially those with lenders, institutional investors, or acquisition ambitions. If you want outsiders to trust your numbers, you need controls that aren't personality-based.

A control is not a policy document. A control is a habit embedded in a process. Examples that actually matter:

- Segregation of duties in payments (the person who sets up a vendor cannot be the person who releases cash).
- Revenue recognition review for non-standard contracts.
- Reconciliations that are prepared, reviewed, and evidenced—not "done in someone's head."
- Access controls in systems that prevent back-dating or altering posted transactions without audit trails.

The leadership move is to frame controls as *how we move faster safely*, not as punishment for past mistakes. Controls reduce the need for heroics. They make it possible to delegate confidently. They turn "trust me" into "here's how we know."

Many companies adopt the COSO framework as a recognized way to think about internal control. COSO is useful because it treats control as a system: the control environment (culture), risk assessment (what can go wrong), control activities (the actual checks), information and communication (how issues surface), and monitoring (how you keep it alive). You don't need to memorize the categories. You do need to build a finance function that behaves like them.

The ethical gray zone: where CFOs earn their salary

The hard ethical moments rarely look like fraud. They look like a reasonable person asking for a reasonable favor at the worst possible time.

A few common gray-zone scenes:

The quarter-end phone call. A business leader asks if you can "recognize it now" because the customer is committed, even though acceptance hasn't happened. The pressure isn't always selfish; it may come from fear—fear of missing targets, fear of layoffs, fear of losing investor confidence.

The "non-recurring" label. Teams want to classify costs as one-time so the adjusted metrics look better. Sometimes it's fair. Sometimes it becomes a landfill where recurring reality goes to die.

The optimistic forecast. A leader insists pipeline will close because it "always does," and they want finance to reflect that confidence. The ethical risk isn't that optimism exists; it's that the forecast becomes a political document rather than an instrument.

In these moments, the CFO's power is not moral superiority. It's process. The CFO can say: "We can't do that without evidence that meets our policy." Or: "If we adjust this metric, we'll document the logic and apply it consistently." Or: "We'll run scenarios—base, upside, downside—and commit to triggers." The gray zone shrinks when rules are explicit, evidence is required, and exceptions are documented.

A simple ethics code your team can actually use

Ethics programs fail when they read like legal disclaimers. Your team needs an operational code they can recall when pressured. A widely used framing in management accounting ethics is: *competence, confidentiality, integrity, credibility*. Translated into plain language:

- *Competence*: don't guess when the stakes are high; ask for help; do the work.
- *Confidentiality*: protect sensitive information; don't use it as currency.
- *Integrity*: don't bend rules for convenience; don't participate in misleading reporting.
- *Credibility*: communicate fairly; disclose limitations; correct errors quickly.

This becomes powerful when paired with a real escalation path. People do the wrong thing most often when they feel trapped—when the only way to say "no" is to personally confront a powerful person without backup. A CFO should make it easy to escalate quietly and safely, and should protect the escalator with visible actions.

AI and modern tools: speed amplifies both excellence and error

Modern finance teams are increasingly deploying automation and AI to accelerate close tasks, categorize transactions, detect anomalies, and draft narratives. The leadership and ethics risk is not hypothetical: if an AI-assisted workflow produces an incorrect entry at scale, you can create a very efficient disaster.

The appropriate stance is neither fear nor blind enthusiasm. It's governance. The same instincts you apply to ICFR apply to AI-enabled processes:

- Define what the tool is allowed to do without human approval (and what requires review).
- Maintain auditability: you should be able to explain why a recommendation was accepted.
- Monitor drift: models and rules can degrade as the business changes (new products, new vendors, new geographies).
- Keep accountability human: the tool can assist, but responsibility doesn't delegate.

If you want a single sentence your team can use: *automation moves fast, so review must be designed in, not added later.*

The CFO as the ethical conscience: a job description in behaviors

"Ethical conscience" can sound lofty until you translate it into behaviors that show up on a calendar:

- You ask for evidence when results look too smooth.

- You make it safe to surface control weaknesses early.
- You insist on documentation for exceptions and non-standard treatments.
- You protect the boundary between owning numbers (accounting) and explaining numbers (FP&A).
- You treat data definitions as commitments, not preferences.

This isn't about being rigid. It's about being consistent. Consistency is what makes a finance function trusted. Trust is what makes a finance function influential. Influence is what makes a finance function valuable.

Actionable takeaway: run an ``Ethics and Control Pre-Mortem'' with your leadership team

Schedule a 45-minute working session with your controller, FP&A lead, and one operational leader you trust. The agenda is simple: assume you are sitting in a board meeting one year from now explaining a reporting failure. Work backward.

Write down:

- The *three most likely* ways reporting could fail in your company (examples: revenue cutoff errors, duplicate payments/fraud, inconsistent KPI definitions, manual journal entry risk, weak access controls).
- For each, name the *owner* (accountable leader) and the *control* that should catch it (what happens, how often, and who reviews it).
- Define one *escalation sentence* any finance team member can use without fear, such as: *"I'm not comfortable proceeding without documented approval under our policy—can we loop in ___?"*

Then pick one control to strengthen immediately—something concrete enough to complete this month. A good choice is often payments governance (vendor setup + bank change approvals) or a clean separation of duties around journal entries and reconciliations. When the team sees one ethical principle translated into one operational upgrade, "tone at the top" stops being a slogan and starts being the way work gets done.

www.ingramcontent.com/pod-product-compliance
Lightning Source LLC
LaVergne TN
LVHW080531210926
848481LV00021B/1711

* 9 7 9 8 8 9 6 6 5 3 0 4 2 *